Images of Modern America

FRIENDSHIP BOTANIC GARDENS

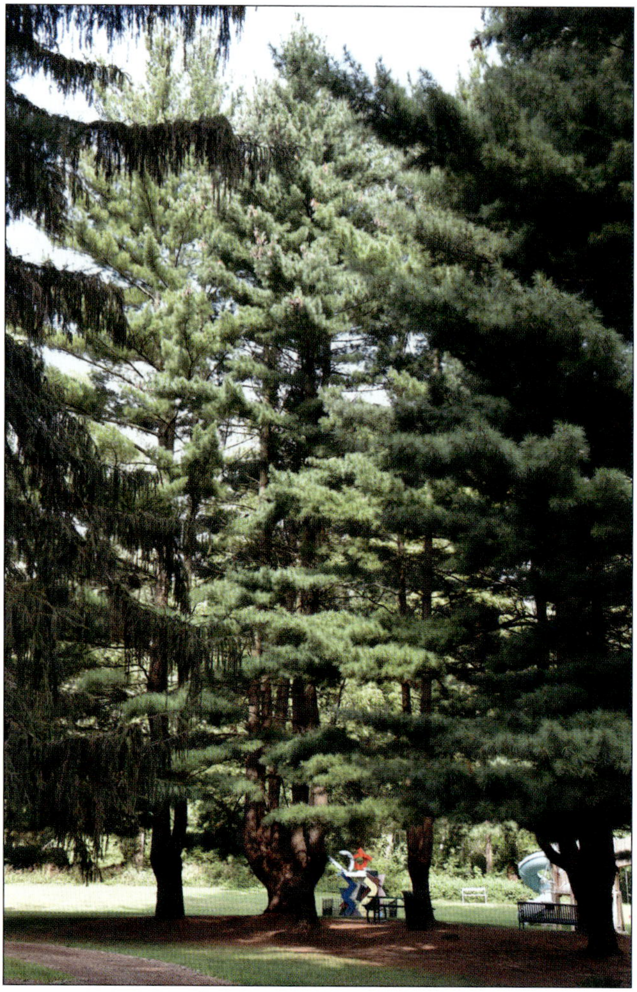

Tall spruce and pine trees are among the outstanding features of Friendship Botanic Gardens. Some of them were planted in the 1930s by the gardens' founders, and some are even older. Their height can be calculated by comparison with a sculpture (lower center), which appears to be a miniature but is actually more than eight feet tall. The sculpture was done by Joe Dahm, former resident of Michigan City, Indiana. (Photograph by Phil Eichas; courtesy of Friendship Botanic Gardens.)

Front Cover: Taken by Phil Eichas, staff photographer for Friendship Botanic Gardens, this image shows a water nymph attributed to Egyptian sculptor Mustafa Naguib around 1975. The figure is nestled among native grasses and daylilies near Lake Lucerne in the Friendship Botanic Gardens in Michigan City, Indiana. It is one of many features recalling the international character of the original gardens, which were founded in the 1930s. (Photograph by Phil Eichas; courtesy of Friendship Botanic Gardens.)

Upper Back Cover: The rose gardens have always been the most spectacular display at Friendship Botanic Gardens. All color photographs were taken by Phil Eichas around 2015.

Lower Back Cover (from left to right): Flowering shrubs attract butterflies. The reflecting pool has served as a focal point for sculpture (center) and for ladies in evening gowns (right).

Images of Modern America

FRIENDSHIP BOTANIC GARDENS

BARBARA STODOLA
WITH FRIENDSHIP BOTANIC GARDENS
FOREWORD BY JOHN LEINWEBER

ARCADIA PUBLISHING

ISBN 978-1-4671-2838-4

Published by Arcadia Publishing
Charleston, South Carolina

Printed in the United States of America

Library of Congress Control Number: 2017963157

For all general information, please contact Arcadia Publishing:
Telephone 843-853-2070
Fax 843-853-0044
E-mail sales@arcadiapublishing.com
For customer service and orders:
Toll-Free 1-888-313-2665

Visit us on the Internet at www.arcadiapublishing.com

Contents

Foreword

From 1934 through the Great Depression, World War II, and the peace of the 1950s to the new millennium, the gardens with their tall trees and lush banks of native shrubbery have provided a tranquil retreat for all who visit.

Friendship Botanic Gardens has a rich history as a cultural cornerstone that enchanted visitors from around the world. Sadly, in the decades since their heyday, the gardens slipped from the public consciousness. In 2014, an ambitious team of community leaders launched a campaign to reintroduce the world to the potential lying dormant within the historic grounds. The new board of directors embarked on a journey to revitalize and transform the gardens into a powerful force for connecting people with nature.

Our mosaic of habitats and lovingly cultivated gardens provide idyllic settings for weddings, concerts, and educational programs. Friendship Botanic Gardens is poised to emerge as our community's defining cultural attraction. This is the dream the new board of directors is bringing to life, and the results so far have been nothing short of spectacular.

The public's interest in the gardens has skyrocketed to unprecedented levels. Thanks to renewed fundraising efforts and the support of diverse corporate and individual donors, Friendship Botanic Gardens is blossoming into a premier destination point for northwest Indiana and beyond. For the first time, the gardens have a paid staff to maintain the grounds and coordinate fundraising efforts.

Community leaders are creating additional gardens in formerly empty spaces. This list now includes a new entryway, Cancer Survivors Garden of Love and Hope, Juvenile Diabetes Garden, Horizon Bank Labyrinth Garden, Irish Heritage Garden, ArcelorMittal Children's Garden, Lithuanian Garden, Rotary Garden, the Native American Garden, the Symphony Garden, and a garden commemorating Veterans of Foreign Wars.

In 2017, more than 25,000 people will experience the gardens, and we look forward to hosting over 50,000 annual visitors in the years ahead for our educational programs, fishing in Trail Creek, the Easter egg hunt, Haunted Trails, and the Turkey Walk and Scavenger Hunt.

Friendship Botanic Gardens are not complete nor will they ever be, for they will weather the uncertainties of the years to come, fulfilling the needs of future generations. Our mission is to provide a space to enjoy nature's splendor in the spirit of tranquility and friendship.

I would like to thank Barbara Stodola and Dr. Rima Binder for researching our past and presenting our future. A hidden jewel is reborn!

—John Leinweber
President, Board of Directors

Acknowledgments

An author who undertakes a book about gardens soon discovers that, in the greater scheme of things, the attention required to produce a book is far exceeded by the daily needs of the gardens. I am doubly thankful to all those who assisted me, now that I realize how dedicated they are to the development of Friendship Botanic Gardens, and how much time they devote to this endeavor. First, I want to acknowledge Dr. Rima Binder, a lover of history and of scholarship, who initiated and supported this project throughout its various stages, and to John Leinweber, current board president of Friendship Botanic Gardens, whose personal contributions and leadership are applauded by his colleagues. This book would not have happened without their encouragement and the approval of their board of directors.

Dozens of volunteers (hundreds, over the years) have adopted these gardens and kept them going. I am mentioning those who supplied me with photographs and information that survived the passage of time: Jeanne Houck and her husband, Dr. Richard Houck, who kept copious records of the many years she was in charge; Alex Rakowski, who assembled and bound thousands of historical photographs; his wife, Jude Rakowski, who recreates gardens and events from a period preceding pioneer settlement; and board attorney Mary Fox, who supplied original documents. For other information, I am thankful to Elta Ehret, Susan Bagby, Ann Dahm, Kevin Firme, Marjorie Crawford, landscape architect Stuart Franzen, Johnny Stimley, Nancy Jahnel-Barnes, Ott-Haverstock and Geisen-Carlisle funeral homes, and the Michigan City Public Library.

Staff photographer Phil Eichas deserves credit for the beautiful color photographs of Friendship Botanic Gardens. For additional photographs, I thank Marianne Olsaker, Jeanette Neagu, Deanna Caulfield, John Rozinski, and Gene Tylisz.

Business manager Triecia Roberts and Ronald Taylor, communications director, generously allowed me working space and access to Friendship Botanic Gardens files. My computer skills were updated by my daughters Madeleine Stodola and Patty Stodola and by Denise Holmes, who also steered me over the rough spots. And finally, I thank Arcadia Publishing and my editor, Stacia Bannerman.

Unless otherwise noted, all photographs are courtesy of Friendship Botanic Gardens. All photographs in chapter six are by staff photographer Phil Eichas, unless otherwise specified.

INTRODUCTION

Friendship Botanic Gardens have a time-honored story to tell. They began as a dream, outlived their founders, survived against all odds, and emerged with a new identity.

The original concept was International Friendship Gardens, a plea for world peace based on the universal love of flowers. The developers were three young men raised on a farm in Wakarusa, Indiana—Joseph, Virgil, and Clarence Stauffer—who combined their wits and resources to enter an ambitious horticultural exhibit in the Chicago world's fair of 1933–1934.

As luck would have it, 1933 not only marked the city of Chicago's 100-year anniversary; it was also the centennial year of Michigan City, which in the 19th century had rivaled Chicago for shipping prominence on the Great Lakes. This coincidence was noted by Phil T. Sprague, civic director of the Michigan City Chamber of Commerce. As a community leader, Sprague was a colleague of Dr. Frank Warren, a physician and real estate developer who owned a few hundred acres at the eastern edge of Michigan City.

After the Chicago world's fair closed, in 1934 the Stauffers were searching for a permanent home for their International Friendship Gardens, and the chamber of commerce helped to convince Dr. Warren that this type of attraction would be an economic stimulus for their town. The collaboration between civic and commercial interests was instituted and repeated, off and on, throughout the tenuous history of the gardens.

Michigan City celebrated its centennial on the Fourth of July weekend in 1933 with a pageant portraying Indian braves and maidens. Eight years later, International Friendship Gardens followed suit with an Indian pageant written by Nola Nance Oliver and titled *Passing of the Pottawattomies*. Sprague himself played the part of an old settler, along with former mayor Martin Krueger, Dr. F.V. Martin, and First National Bank founder Walter W. Vail. Joseph Stauffer had arranged for an authentic Indian powwow to be staged by full-blooded Indians, including Winnebago chief Whirling Thunder, for this spectacular performance.

The Native American Heritage Days, reestablished in Michigan City a few years ago and currently organized by Jude Rakowski, can draw on this long and distinguished local history.

When the Stauffers acquired the property, they enhanced the native woodlands by adding more trees—200 Norway spruce from a nearby nursery that was going out of business. Some of these spruce trees, now 70 feet tall, can be seen alongside the children's garden. Other trees are even older, some dating to the Civil War era. Stuart Franzen, a landscape architect and board member, points out that the land has never been farmed and a steady source of water is provided by Trail Creek, which runs through the 105-acre grounds. The existence of legacy trees, Franzen says, is what distinguishes this site from other botanic gardens in the Midwest.

With the assistance of local garden clubs and proud ethnic organizations, the Stauffers had 14 ethnically themed gardens for their opening celebration in 1936. One garden was Chinese, another Persian, and the remainder were based on western and central European countries. Over the years, these specially designed gardens increased in number, as did the persons who came to visit. International Friendship Gardens became famous and were featured in newspapers and national magazines. In 1957, the Voice of America produced a 600-word radio script describing International Friendship Gardens and made it available in 43 different languages for listeners all over the world.

But times changed and resources dwindled. Michigan City is a town whose fortunes have waxed and waned, and International Friendship Gardens lost their stellar position among the local priorities. More than once, the gardens were rescued by volunteers who simply fell in love with their astounding natural beauty. In a world that became increasingly noisier and more hectic, here was a tranquil spot that recalled the past and possibly foretold the future.

Jeanne Houck and her husband, Dr. Richard Houck, arrived on the scene in 1970 and did their best to keep International Friendship Gardens afloat for more than 40 years.

During that time, Michigan City's fortunes rebounded. Lighthouse Place, an upscale shopping outlet mall, was established on land unused since the closing of the Barker family's railroad car industry. Lighthouse Place developers joined forces with La Porte professional leaders, and boosted by a challenge grant from the Lilly Foundation of Indianapolis, they established the Unity Foundation of La Porte County—soon to become a major funding source for local not-for-profit organizations. A few years later, Blue Chip Casino opened its riverboat, coincidentally on the same Trail Creek that flows through Friendship Gardens, a few miles farther upstream. The city entered into an agreement to accept funding from Blue Chip Casino for the betterment of Michigan City—an arrangement continued today by present mayor Ronald Meers, thus providing major revenues for worthy causes.

When Long Beach resident John Leinweber became president of the board in 2014, he determined to make the gardens a successful venture once again. He found new financial backers. He and his family made substantial contributions to improving the grounds. An 18-person board was assembled, with community leaders, youthful retirees, and persons from all walks of life. Dr. Rima Binder, a university professor of Lithuanian descent, became vice-president and took charge of revitalizing the ethnic gardens. The board of directors determined the need for a name change—to Friendship Botanic Gardens—to indicate a stronger emphasis on educational pursuits, to attract a wider and younger audience, and to signify the new direction in which the gardens are being taken. What they started from—and what they are accomplishing—will unfold in the following chapters.

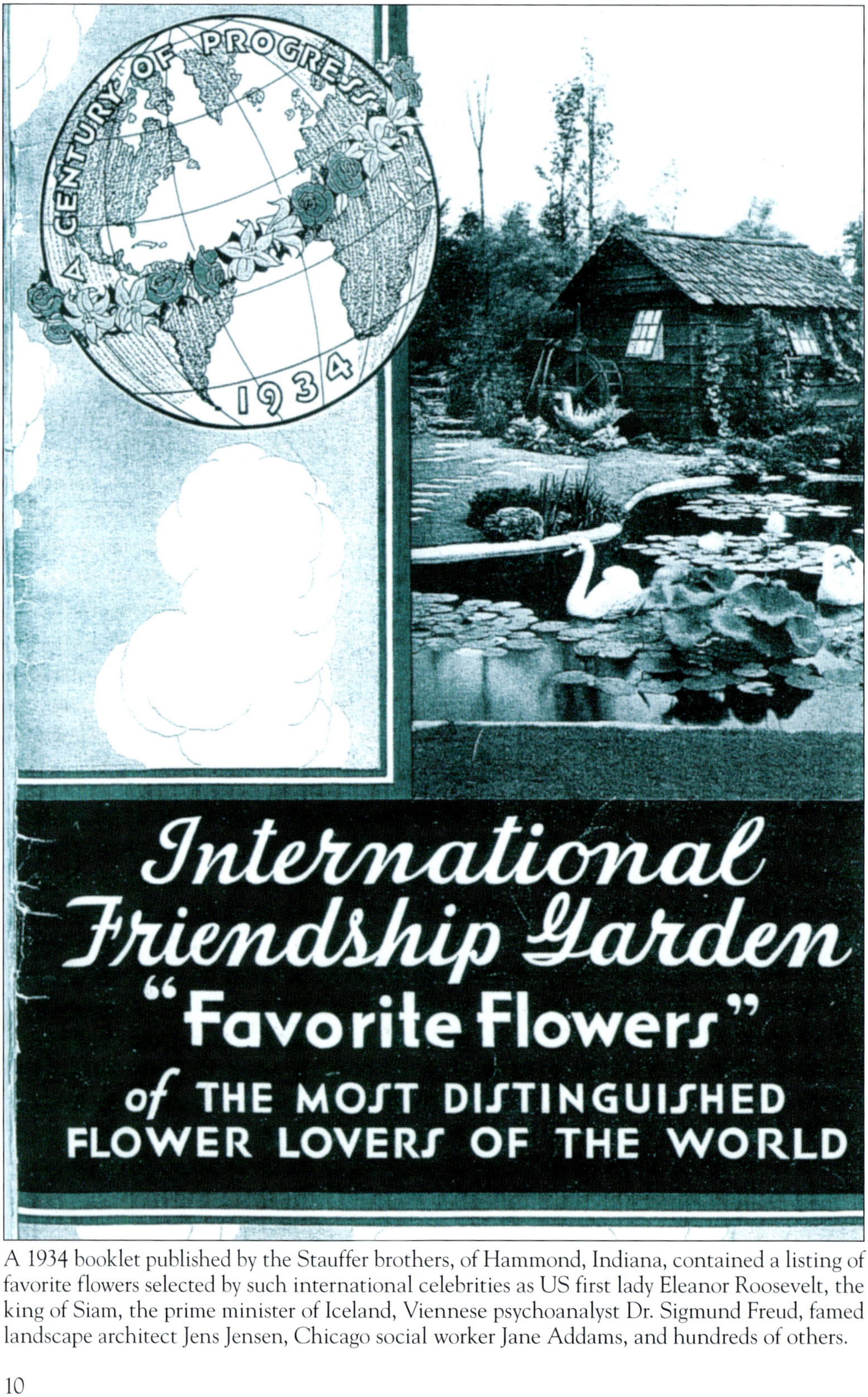

A 1934 booklet published by the Stauffer brothers, of Hammond, Indiana, contained a listing of favorite flowers selected by such international celebrities as US first lady Eleanor Roosevelt, the king of Siam, the prime minister of Iceland, Viennese psychoanalyst Dr. Sigmund Freud, famed landscape architect Jens Jensen, Chicago social worker Jane Addams, and hundreds of others.

One

Century of Progress

When the Chicago world's fair opened in 1933, the country was sunk in its deepest economic depression in 40 years. The Century of Progress Exposition, as the fair's title indicated, tilted toward the future and the improvements promised by scientific and technological advances. But some exhibitors, such as the Stauffer brothers, pinned their message on a romanticized vision of America's fertile fields and homegrown charm. Their Old Mill Garden presented a flower-filled enclave with a wood-shingled cottage, winding paths, welcoming benches, and a gently flowing stream inhabited by a couple of swans. It was a prototype for their future 100-acre garden spot in Michigan City, Indiana.

Although the fair's international components included an English garden, two Italian gardens, and two Japanese gardens, most horticultural exhibits focused on American themes: Suwanee River cabin, Abraham Lincoln's 1820 Indiana home, an Indiana dunes landscape, and the Stauffers' Old Mill Garden. An Old World atmosphere was captured in entertainment areas such as Swiss Village, Old Mexico, and Streets of Paris.

The fair's ambitious organizers encouraged Chicago's ethnic communities to involve cultural and intellectual leaders of their native countries, asking for "the greatest men in every line of science, industry and philosophy." In keeping with this emphasis, the Stauffer brothers, though American born, extended their garden theme to include hundreds of world leaders.

As the fair moved into its second year, the Stauffers introduced an International Friendship Garden and a 32-page booklet listing "Favorite Flowers" of "the most distinguished flower-lovers of the world." Their booklet showed the results of personal correspondence with prominent individuals throughout the world: kings and queens, sultans and maharajahs, prime ministers, Nobel Prize laureates, musicians and authors, scientists, inventors, military officers, celebrities, and leaders from all walks of life, who evidently took the time to answer the Stauffers' query as to what was their favorite flower. Every effort was made to include these favorites in their 1934 garden.

"Who would not be cheered in the midst of a beautiful garden," intoned the booklet's foreword. "It matters not from what land or climate, the love for beauty is much the same."

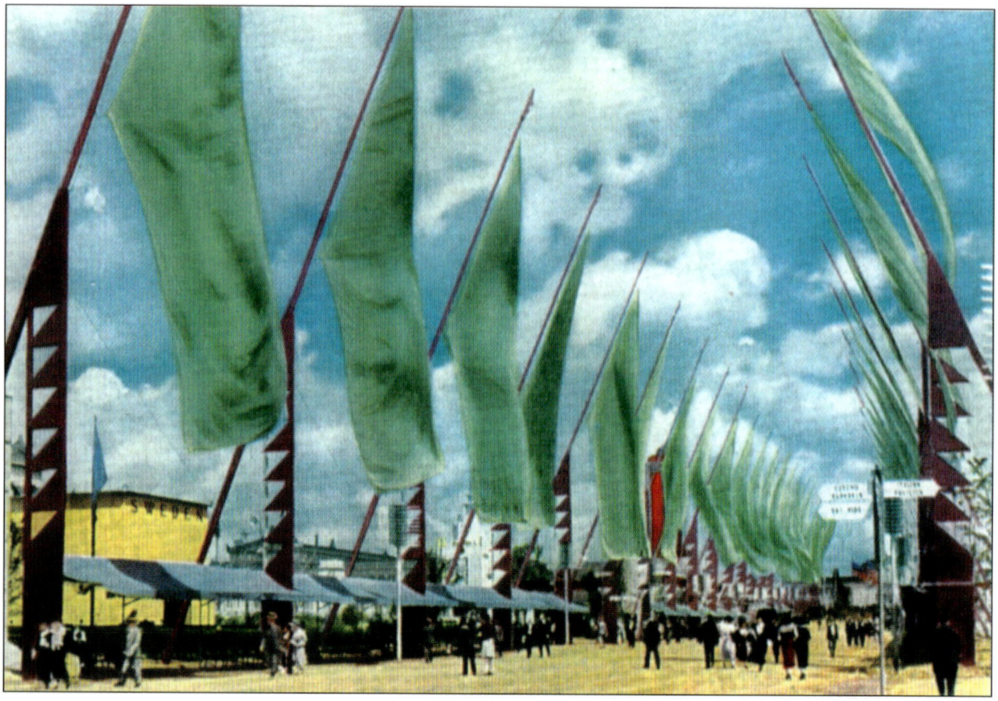

The Avenue of Flags lent an air of grandeur to the Century of Progress Exposition, welcoming visitors to this once-in-a-lifetime event: the Chicago world's fair of 1933.

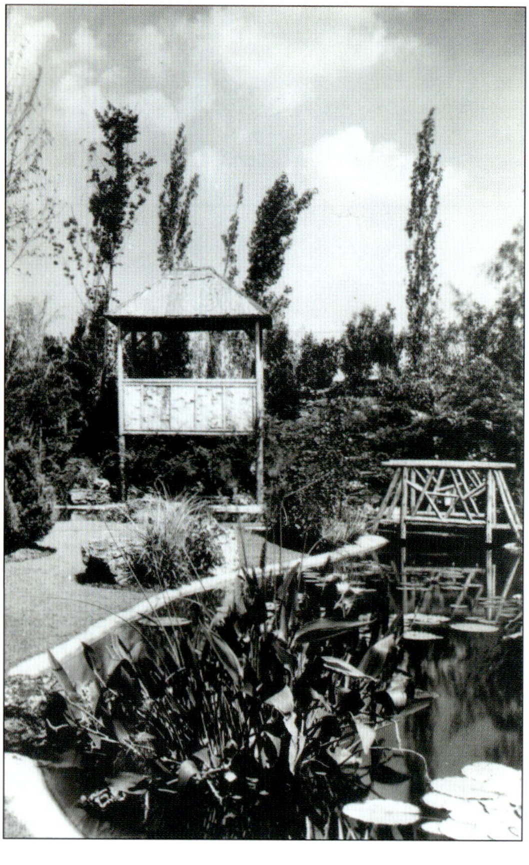

A lookout tower provided a vantage point for fairgoers to have a panoramic view of the Stauffers' Old Mill Garden, with its rustic footbridge crossing over a lily pond.

The tranquil setting of the Old Mill Garden was enhanced by a rock garden, native grasses, and a waterfall. The picturesque garden was designed by landscape architect Charles O. Speicher, of South Bend, Indiana, and photographed for souvenir postcards. (Courtesy of Alex Rakowski.)

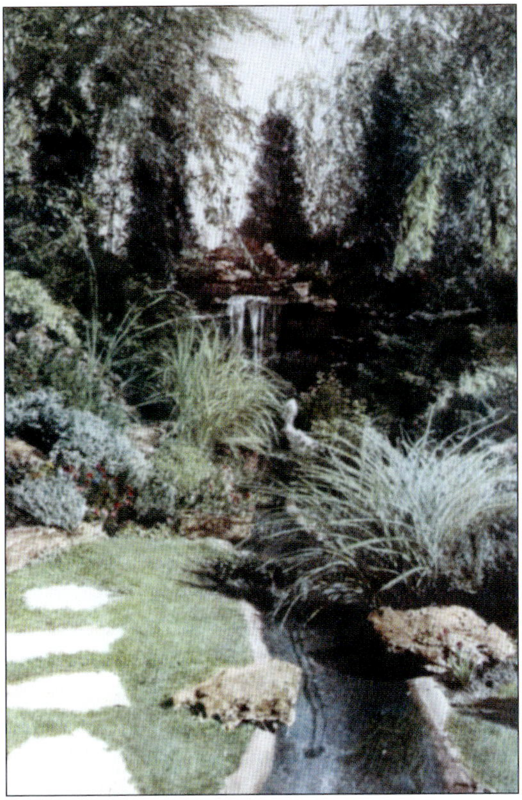

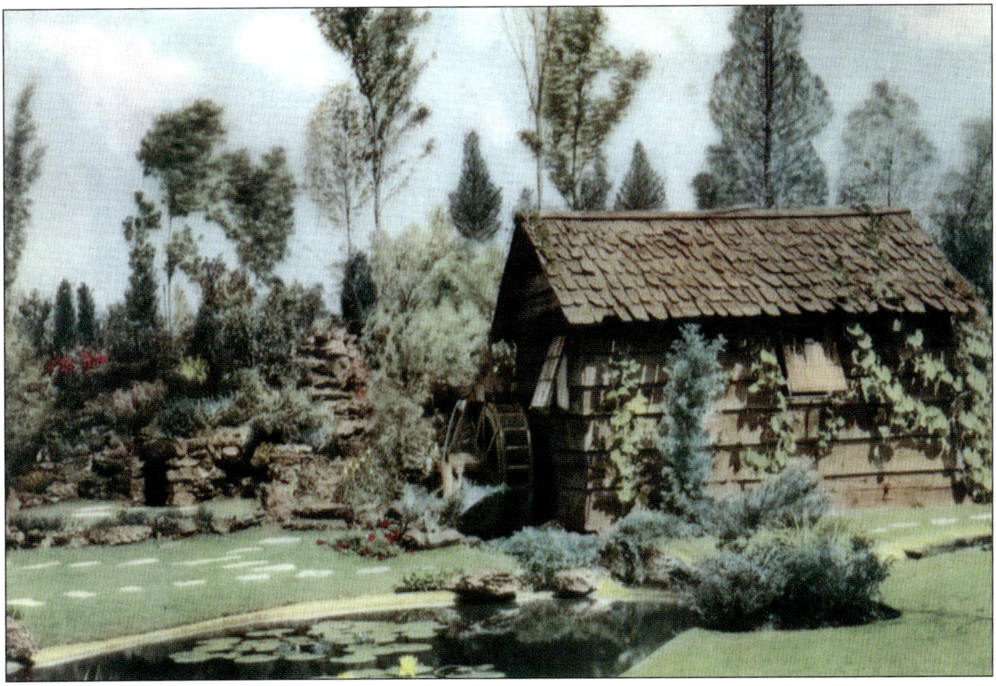

The centerpiece of the Stauffers' horticultural exhibit was an ivy-covered old mill, with a wood-shingled roof and a waterwheel. (Courtesy of Alex Rakowski.)

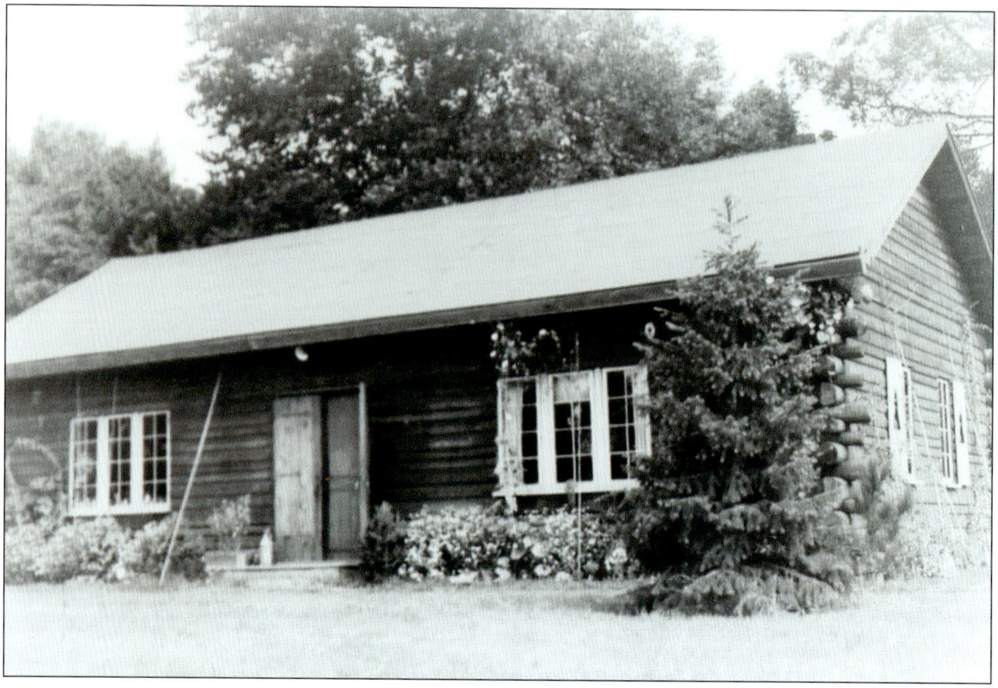

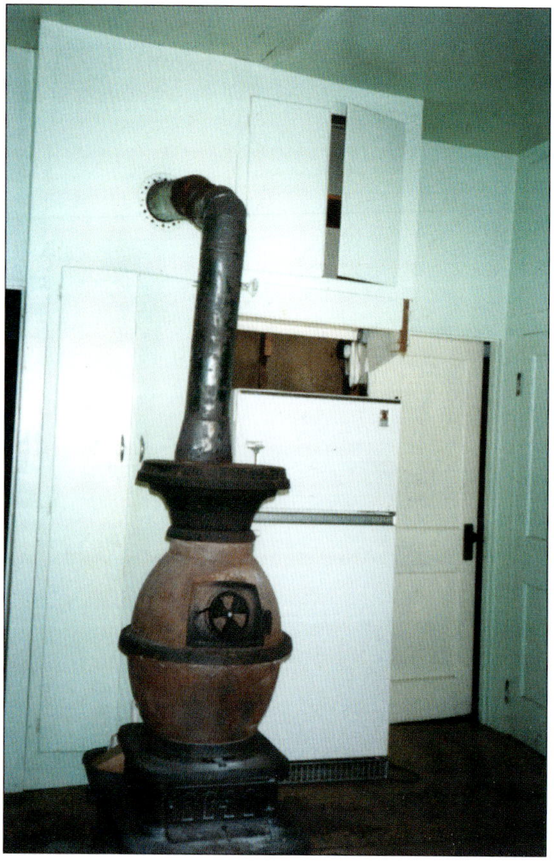

The Alaska cabin (above) exhibited at the Chicago world's fair showed the comfortable though rugged lifestyle in the frontier territory of Alaska. The only source of heat is a dependable wood-burning stove pictured at left. After the fair closed in 1934, the Stauffers made arrangements with US secretary of the interior Harold Ickes to have the building taken apart and shipped to Indiana, where it was reassembled at International Friendship Gardens.

A woman dressed in a late-19th-century costume guided visitors to the Alaska Territory exhibit at the Century of Progress International Exposition.

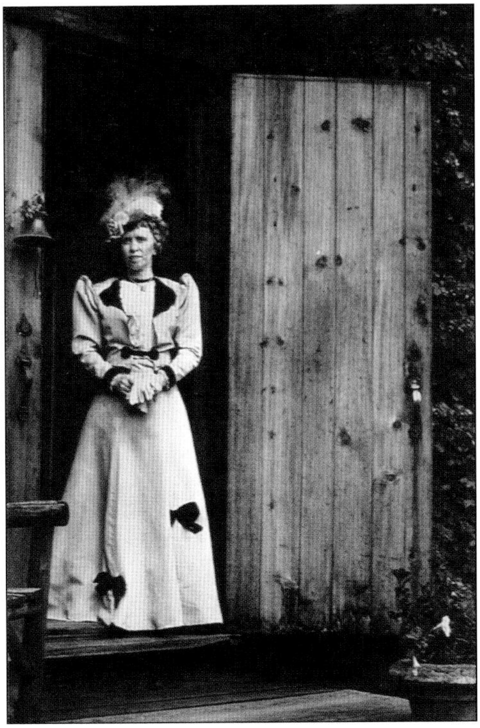

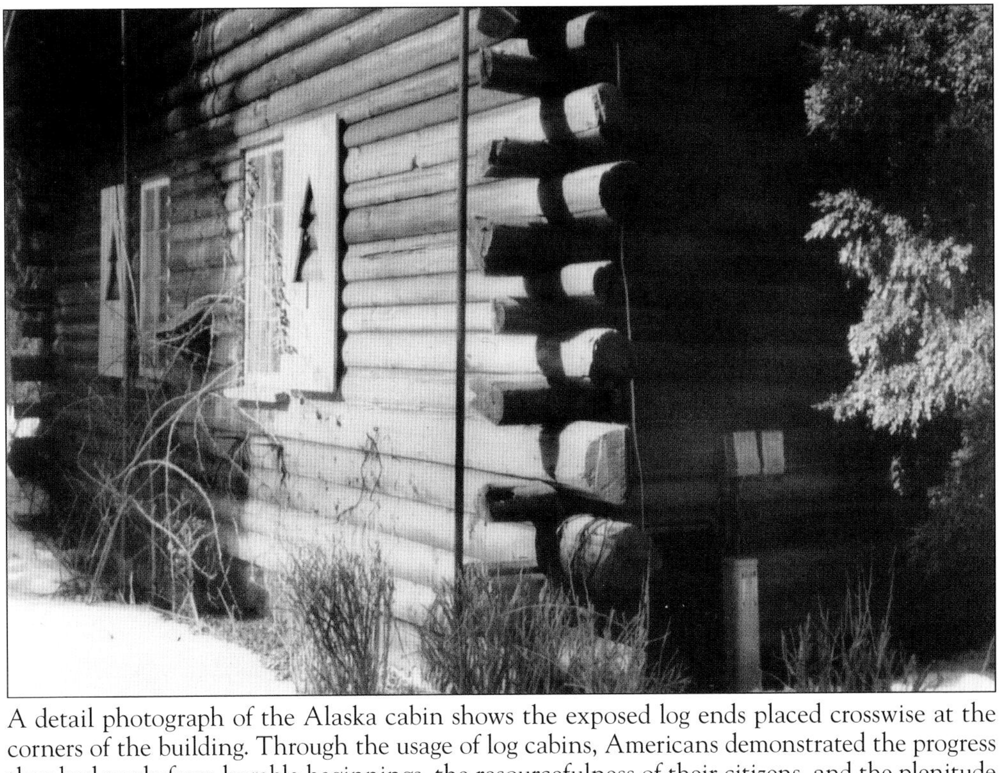

A detail photograph of the Alaska cabin shows the exposed log ends placed crosswise at the corners of the building. Through the usage of log cabins, Americans demonstrated the progress they had made from humble beginnings, the resourcefulness of their citizens, and the plenitude of lumber available in their country.

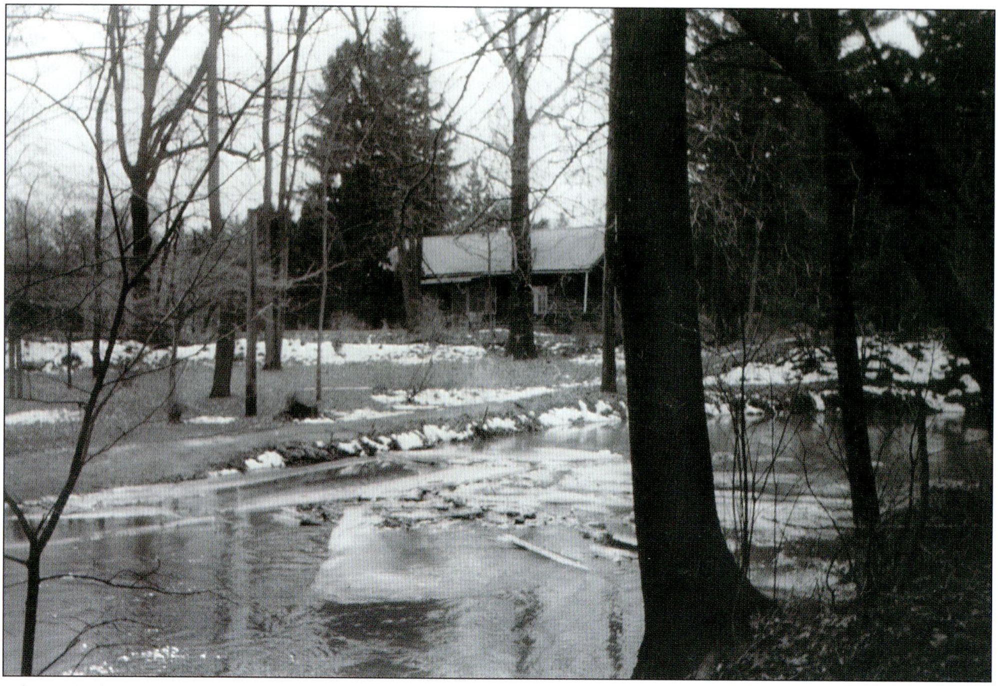

The Alaska cabin, moved to Michigan City from the Century of Progress Exposition, served as home and headquarters for the Stauffer family members during the years that each of them was in charge. Special challenges were presented by seasonal frost, snow, floods, and thaws.

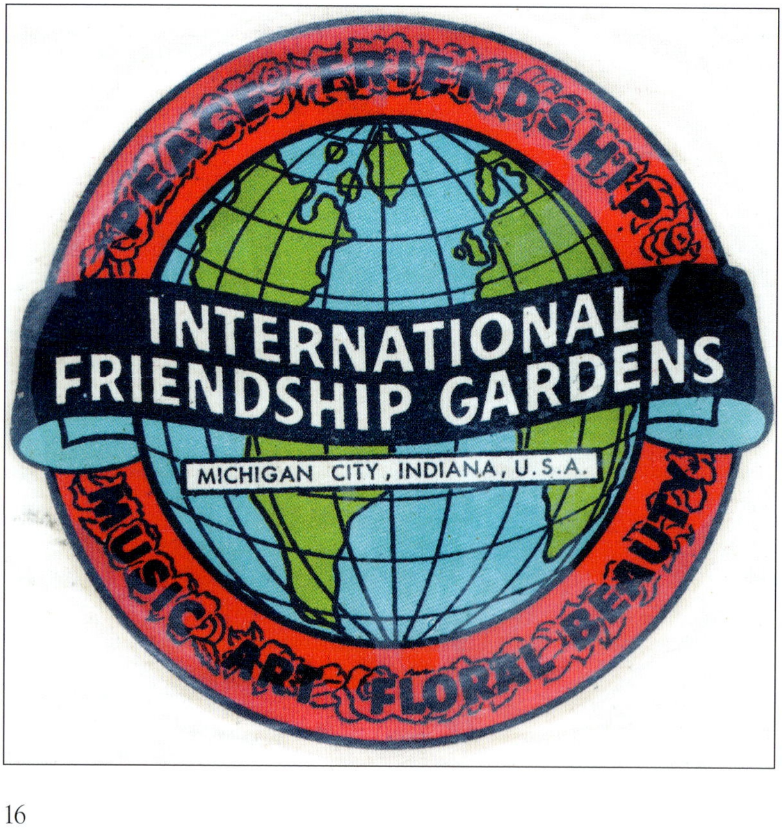

For the International Friendship Gardens they developed in Indiana, the Stauffers had a logo designed that depicted the globe encircled with the motto "Peace Friendship Music Art Floral Beauty."

Two

MOVING TO INDIANA

The Stauffer brothers could trace their ancestry back to the Pennsylvania Quakers, generations of farmers moving consistently westward and arriving in the mid-19th century in the small community of Wakarusa, Indiana. With parents who valued higher education and individual achievement, Virgil studied art and music and Joseph entered Moody Bible School in Chicago but then became a mechanical engineer, with a degree from Armour Technical School. Facing the practicalities of life, they opened a radio and electronics store in Hammond, Indiana. Clarence, the youngest brother, graduated from Purdue University and tried his hand at raising poultry, at real estate, and at inventing various types of machinery. Eventually, Clarence joined his wife managing Citizens Bank of Wakarusa, a business owned by her family. All three brothers loved gardening and were devoted to the cause of world peace.

In 1926, Clarence and a friend created a floral display to mark the opening of the Chicago Merchandise Mart. This experience was the prelude to the Old Mill Garden at the Century of Progress Exhibition, which involved the collaboration of all three brothers, plus generous support from the Citizens Bank of Wakarusa. When the world's fair closed in 1934, the brothers searched for a permanent home for the gardens they envisioned as a testament to "peace and goodwill among all peoples of the world."

Dr. Frank Warren, a Michigan City doctor and real estate developer, was impressed by his visit to the Chicago world's fair—so greatly impressed that he turned over 100 acres of land to the Stauffer brothers for their continuance of gardens representing many countries of the world. The Michigan City Chamber of Commerce helped convince him this donation would bring great economic benefits to the region. The property was adjacent to Michigan City's Memorial Park and to the town of Pottawattamie Park, which Dr. Warren was developing as a residential community.

The 99-year $1 renewable lease was signed, and the Stauffers went to work. According to John H. Gleason, a chronicler of the times, "They pitched in shoveling, loading, ripping up weeds, grubbing, smoothing off humps, transplanting big and little trees, cutting brush, killing sand burrs and dredging the creek, along with the one thousand other things that needed to be done." For the next 50 years, their family labored to keep International Friendship Gardens luxuriant, gloriously blooming, and open to the public.

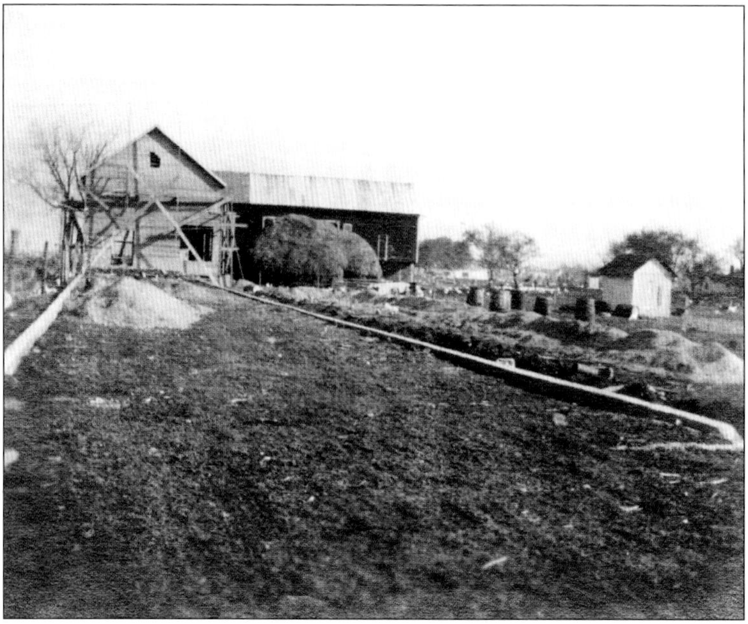

This old photograph depicts the Stauffer family farm in Wakarusa, Indiana, where the brothers' love of gardening was fostered. Both Joseph and Virgil Stauffer went on to run International Friendship Gardens until their deaths in 1957. Their sister Sadie took over until she died in 1964, and Clarence then became the last family member to manage the gardens. (Courtesy of Alex Rakowski.)

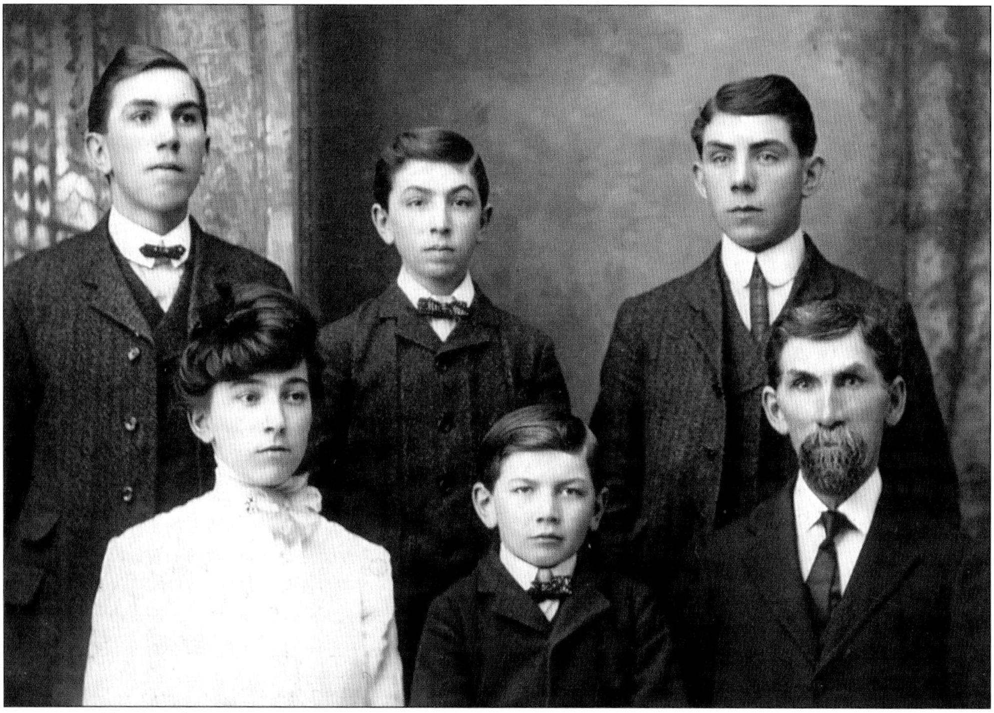

In a Stauffer family portrait taken around 1900, Clarence (first row, center) is seated between his father, John Stauffer, and his sister Sadie, who later married Albert Ehret and moved to Hammond, Indiana. The three older brothers are, from left to right, Boyd, Virgil, and Joseph. All the siblings took turns running International Friendship Gardens except for Boyd, who died in 1936. Their mother had died in 1899, when Clarence was four years old.

Virgil Stauffer was a pacifist and conscientious objector, as well as a horticulturist. A talented vocalist, he studied art and music at Chicago institutes and made his debut as an opera singer. His connections in the Chicago artistic and literary worlds led to the early establishment of classical musical concerts in the gardens.

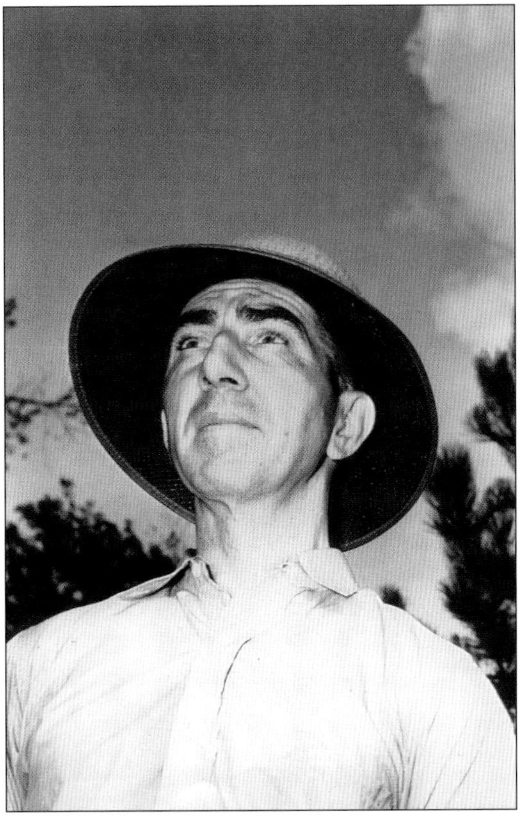

Among the many visitors to the gardens, this group of nuns from Mishawaka, Indiana, posed for a photograph with Joseph Stauffer. After suffering a back injury, Joseph was forced to curtail his gardening activities, but he still continued supervising the irrigation system, carrying on an extensive search for large bells for their property and greeting guests. Although the Stauffer brothers became Christian Scientists, their gardens were an inspiration to persons of many faiths.

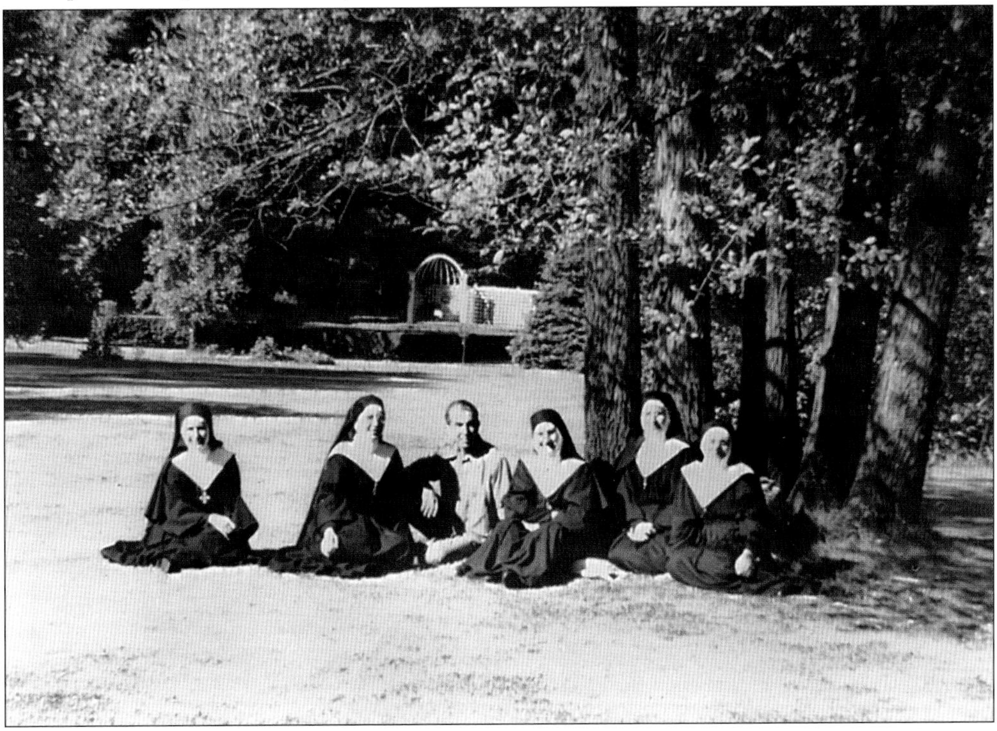

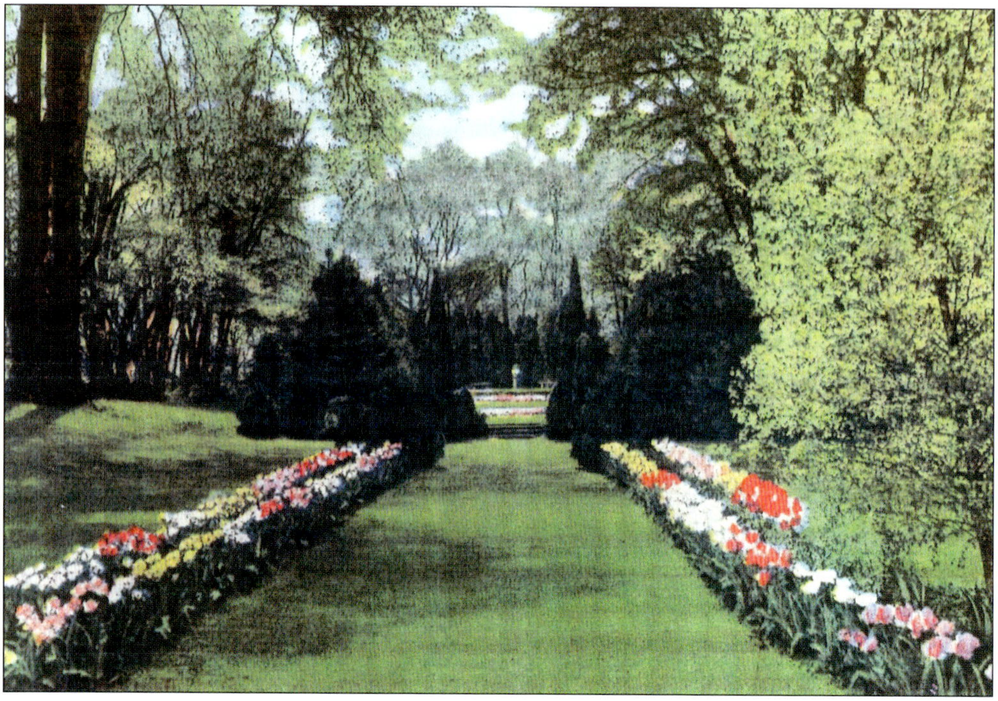

Early photographs of International Friendship Gardens show plantings of the generous gift sent from Queen Wilhelmina and the Bulb Society of Holland—200,000 tulips, lining the pathways (above) and filling flower beds in the Holland garden (below).

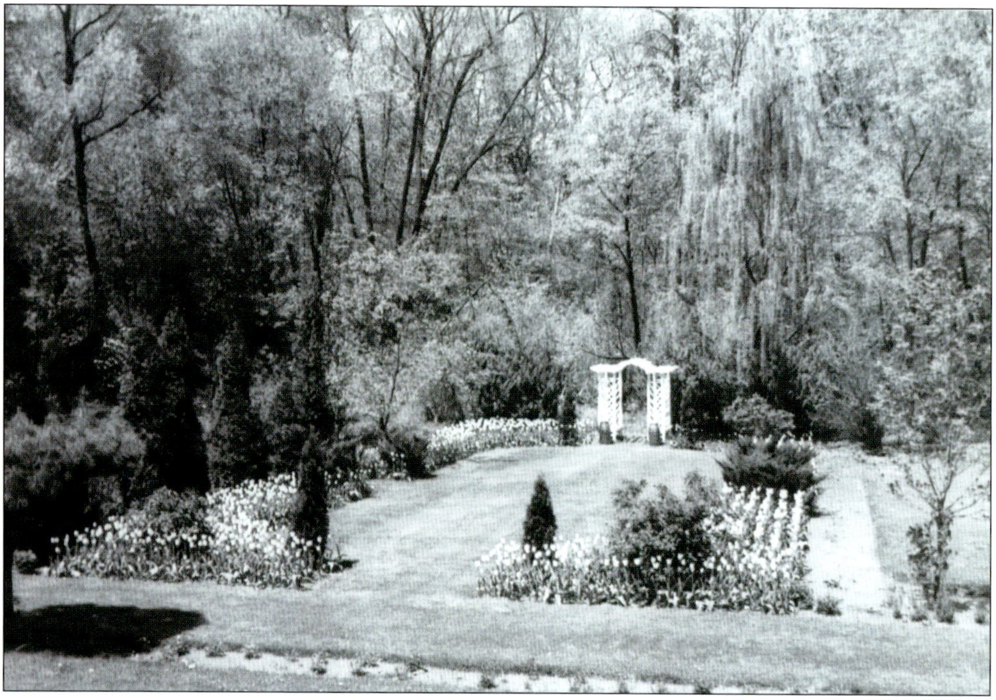

The arched half-moon bridge, a type of bridge that originated in China, became a favorite spot to pose for souvenir snapshots. The Chinese bridge is still in existence and is still a favorite spot.

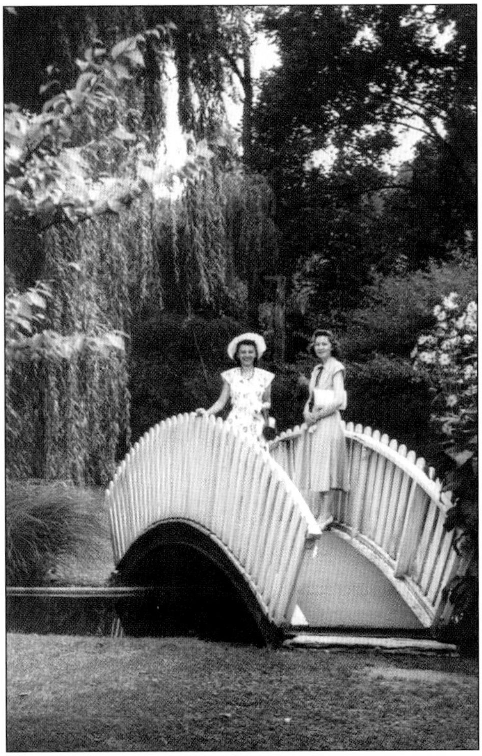

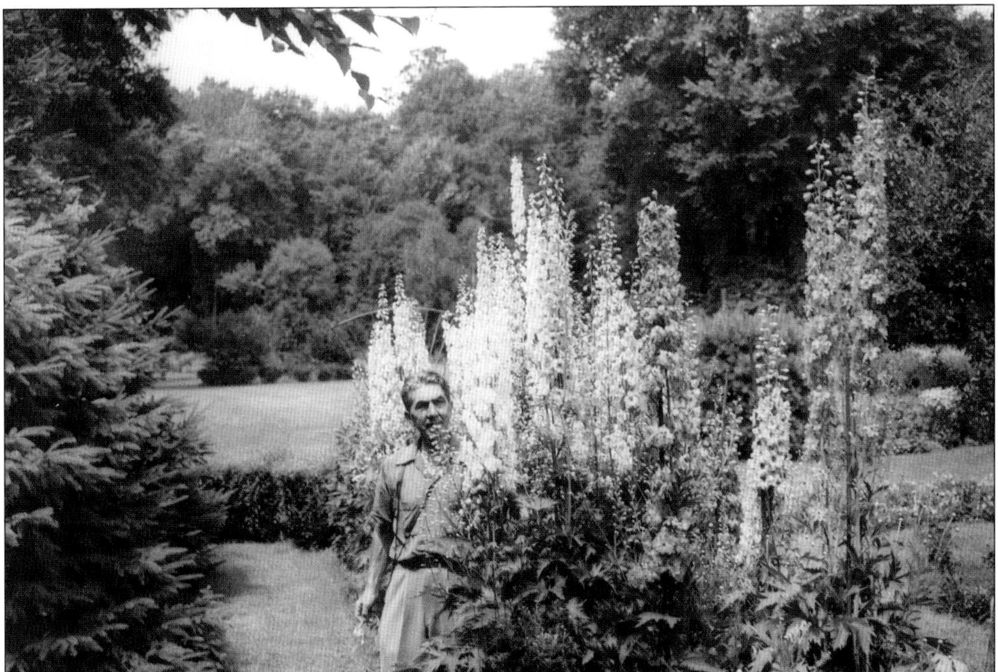

Virgil Stauffer is pictured amid seven-foot-tall delphiniums in the Polish gardens, one of the 14 original ethnic gardens developed in Michigan City. The Polish gardens remain in existence today, although in a different location, with more subdued plantings.

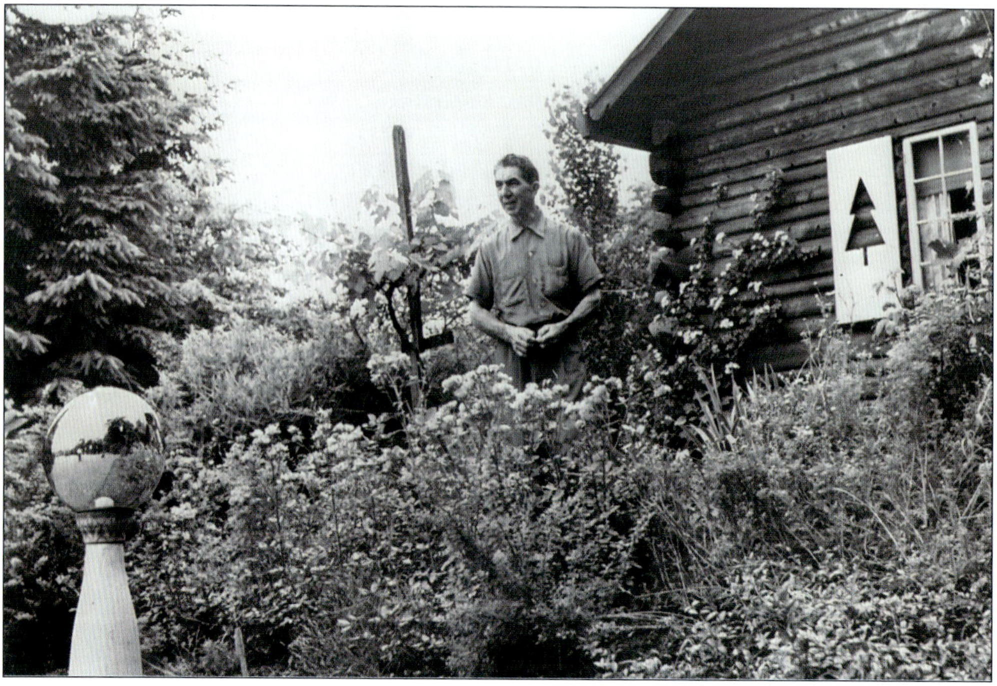

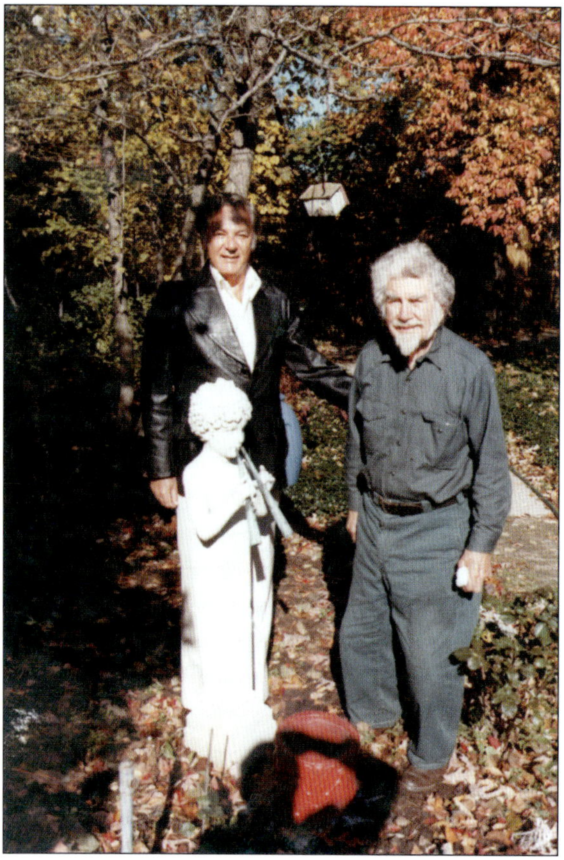

During the glory days of International Friendship Gardens in the late 1940s, Virgil Stauffer stands beside a reflecting ball in his well-tended flower border. Climbing roses adorned the Alaska cabin that functioned as the residence of Virgil and his brother Joseph Stauffer.

In a much later photograph, Clarence Stauffer (right) is pictured with a visitor identified as Dr. Thomas Davis, prime minister of the Cook Islands. The small sculpture of a double-flute player somehow found its way to Michigan City from the 1933 Century of Progress Exhibition. (Courtesy of John Rozinski.)

Clarence Stauffer moved from Wakarusa to Michigan City after his sister's death in 1964. He insisted on maintaining the gardens as a private enterprise, even though the project was overwhelming. Clarence could look back on a lifetime of notable achievements, including the 1931 invention of a gyroplane, which he described to newspapers as a forerunner of the helicopter. He lived to celebrate his 100th birthday in 1986.

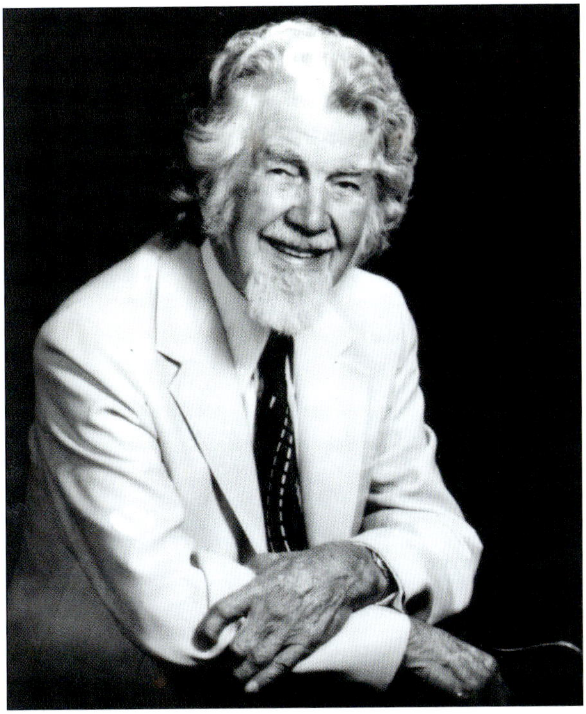

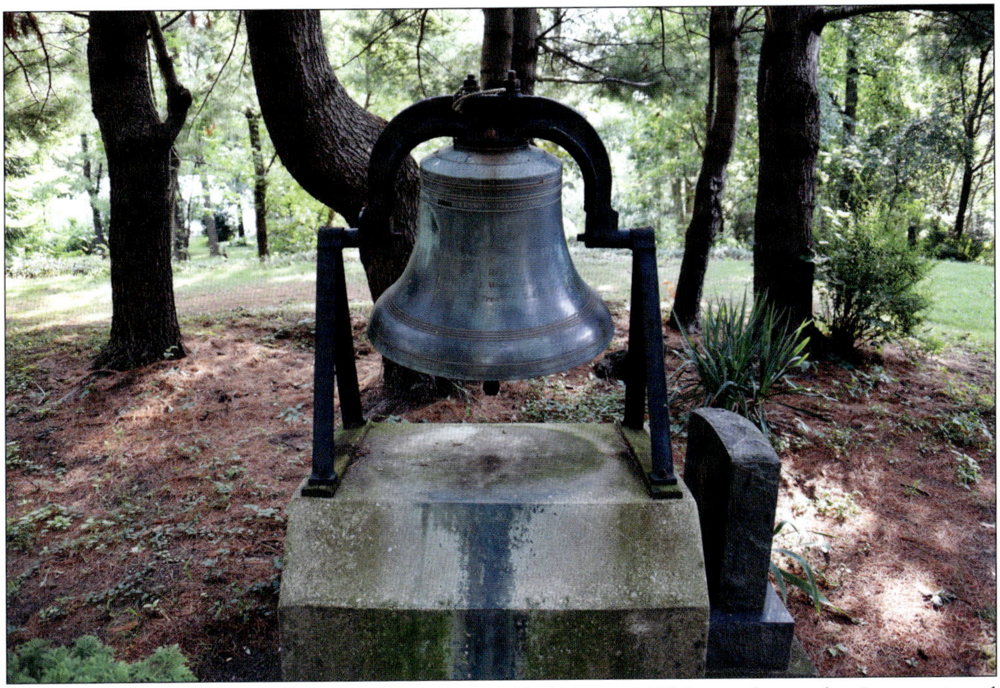

The Elston Elementary School bell, an 800-pound bell cast in 1896, was donated to International Friendship Gardens by Michigan City Schools and dedicated at ceremonies on July 4, 1983. The bell was placed on a specially designed concrete base. An engraved stone marker was later placed beside it, in memory of Clarence Stauffer. He is buried in Wakarusa, Indiana, with other family members.

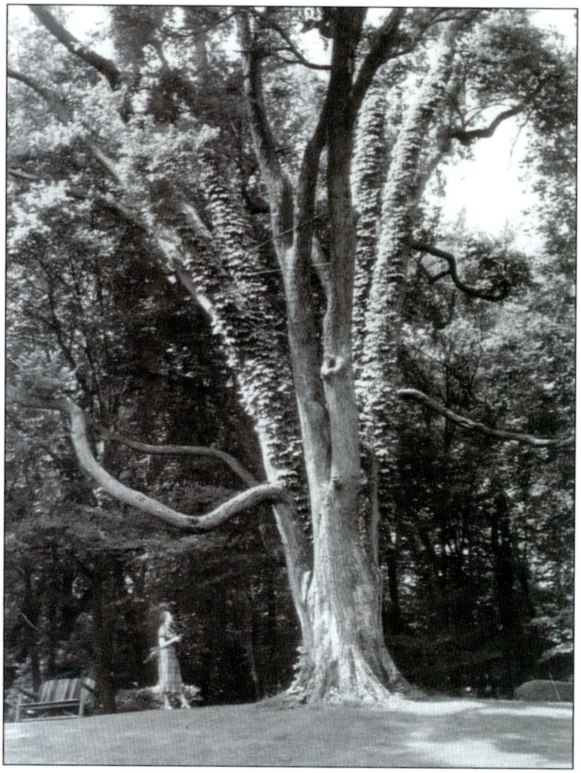

The Great Elm, a site of Indian folklore, stood almost 80 feet tall near the center of International Friendship Gardens. No longer in existence, the great tree's location was indicated on early maps of the area.

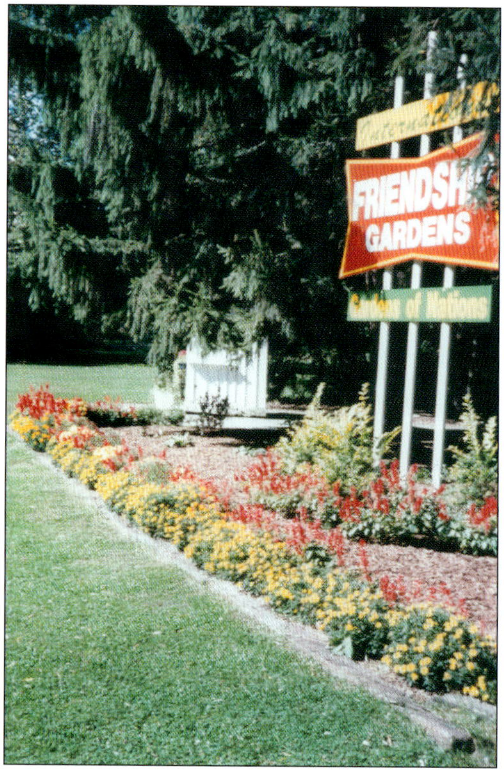

The original entrance sign for International Friendship Gardens, made of painted wood, is still preserved on the grounds, although a new digital LED sign now stands out front.

Three

International Friendship Gardens

With 100 acres of fertile Indiana soil, a valley floor surrounded by old-growth trees and fed by underground springs, and Trail Creek meandering toward Lake Michigan, the site was a dream come true for the Stauffer brothers. The area was steeped in local legend, as the former gathering place for Native Americans and a camping spot in 1675 for French missionary Fr. Jacques Marquette.

The *Michigan City Evening Dispatch* carried the headline story on March 13, 1936, of Dr. Frank Warren's generous gift to establish the gardens. Fifty acres were to be left as natural woodlands, the remaining 50 to be cultivated as gardens. The Michigan City Chamber of Commerce reported the land valued at $150,000.

Following the theme established at world's fair gardens, the Stauffers drew upon their connections with foreign dignitaries and heads of state. King Gustav V sent blueprints of the Swedish royal gardens. Queen Wilhelmina sent 200,000 tulip bulbs from the Netherlands. King George VI dispatched one of his royal gardeners to supervise installation of the English garden. A display of orchids arrived for the opening ceremonies, a gift from Guatemalan president Jorge Obico. Lotus plants came from Egypt. Persian prime minister M.A. Foranghi sent Damascus roses from Teheran. Other early respondents were Hungarian, Lithuanian, and Polish diplomats.

Before any planting could begin, however, the wetlands had to be dredged. Joseph Stauffer, a mechanical engineer, devised a drainage and watering system. A pond was created, suggestively named after Switzerland's Lake Lucerne, with a small island designated as the Theatre of Nations. Another performance area was known as Little Symphony Theatre, and the gardens soon began hosting classical music, Broadway show tunes, ethnic dance groups, and light opera.

Each of the ethnic gardens expressed the character of its native country. The French garden had strictly formal plantings. The Scottish garden had a traditional cradle fence. A half-moon bridge was built to accent the Chinese garden. The Grecian garden included tall, slender junipers arranged with niches for statuary. Columns in the Italian gardens replicated the columns of the ancient Roman Forum. The president of Lithuania traveled to Michigan City to plant a commemorative tree in the Lithuanian garden.

During the 1940s, visitors from all over the United States arrived at International Friendship Gardens and, following World War II, they came from many foreign countries.

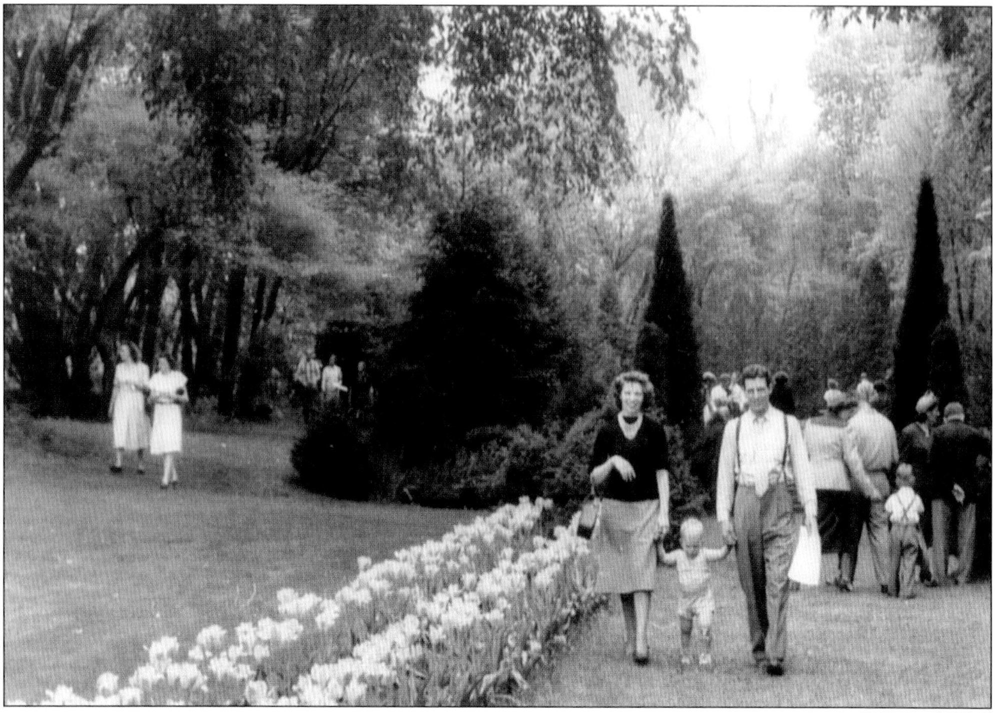

Mother's Day officially opened the season, and throngs of visitors (above and below) arrived at International Friendship Gardens. Groups of families strolled along the tulip-bordered pathways, admiring the reflecting pools and neatly trimmed shrubbery.

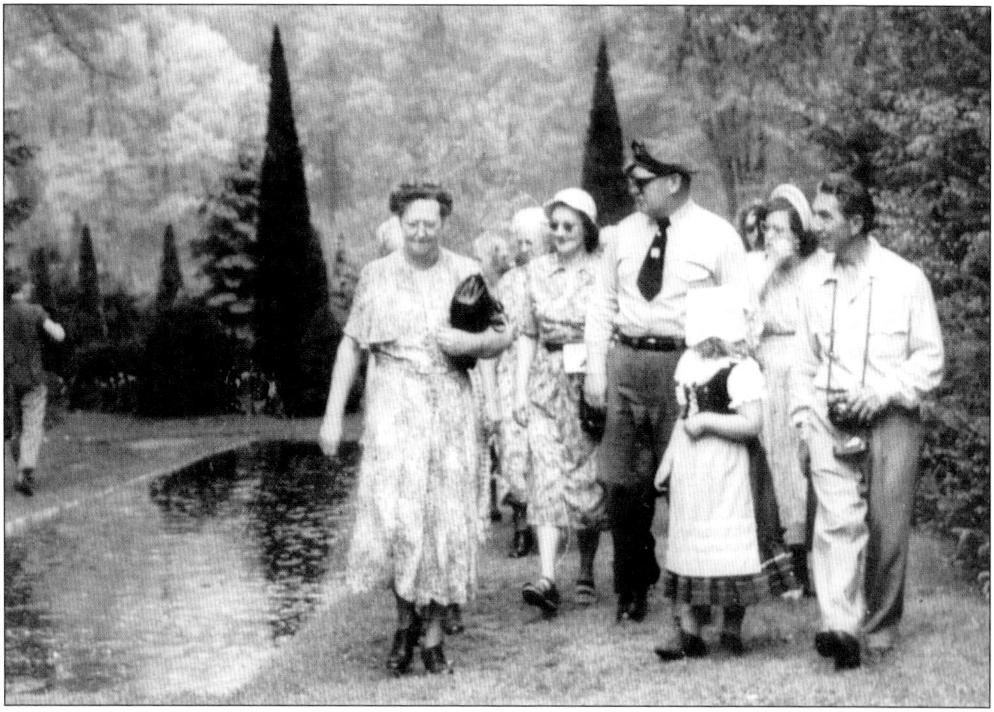

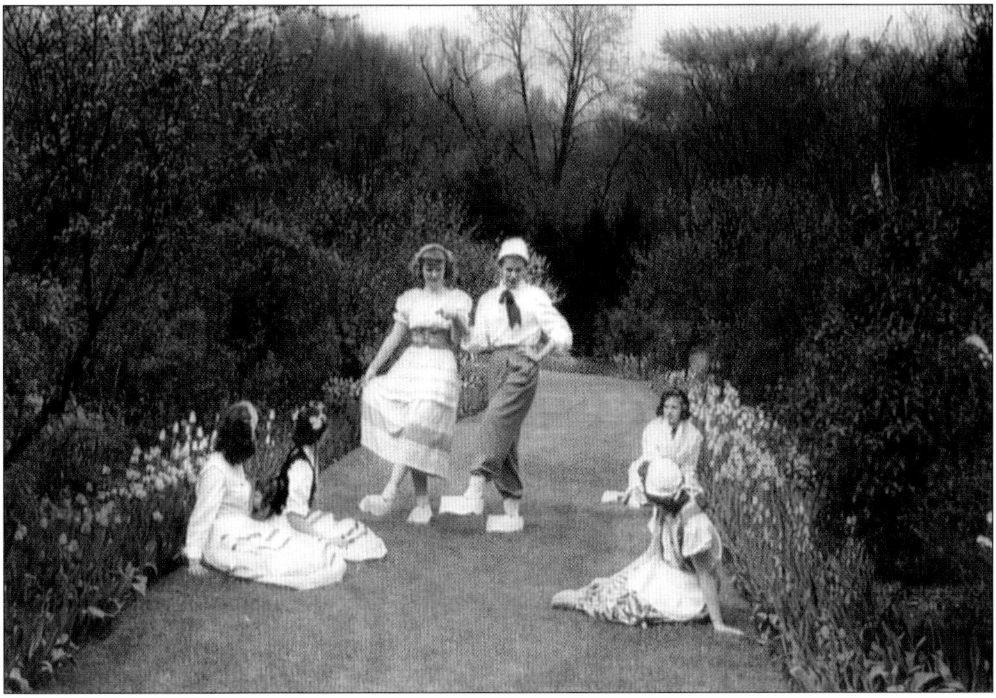

Dutch teenagers, garbed in traditional costumes and wooden shoes, staged an impromptu dance performance.

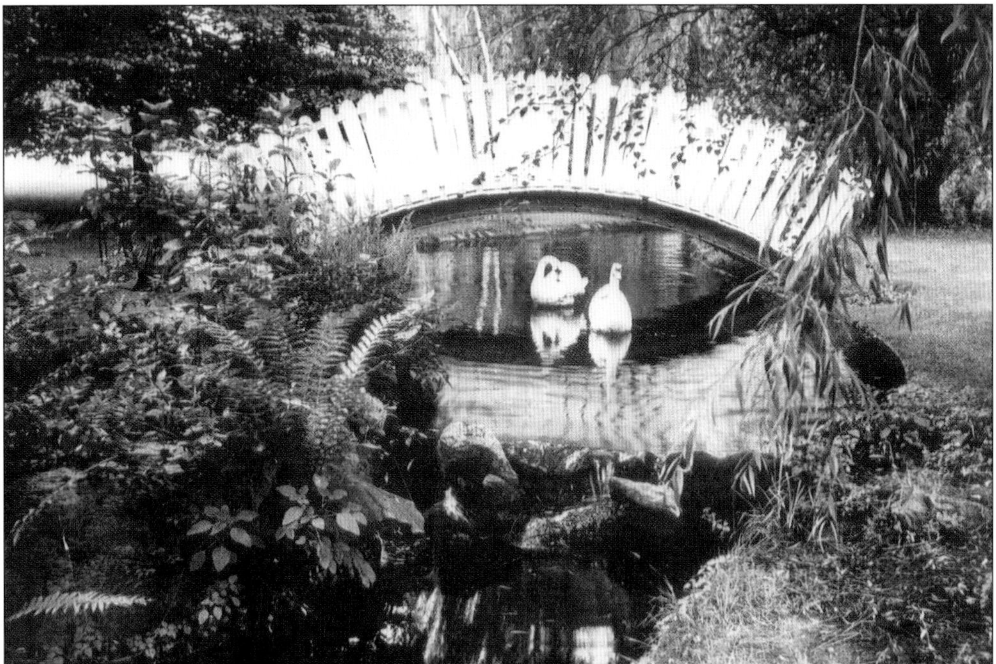

The picturesque character of the half-moon bridge, spanning a little stream in the Chinese garden, was enhanced by a couple of swans (evidently made of plaster and brought to Michigan City from their previous home at the Chicago world's fair). This area was referred to as Willow Plaza.

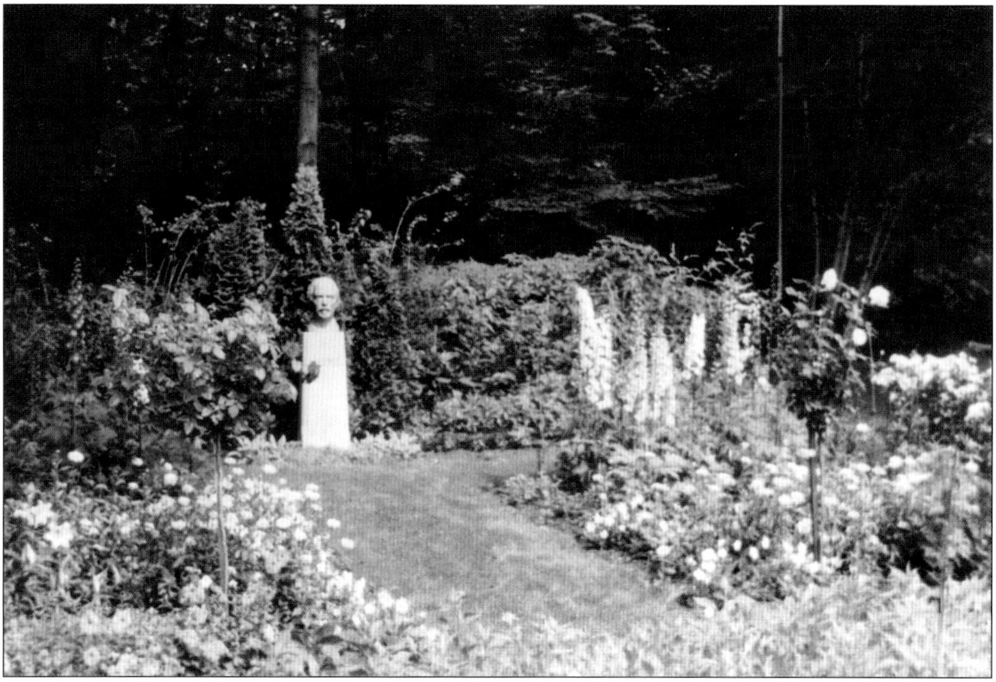

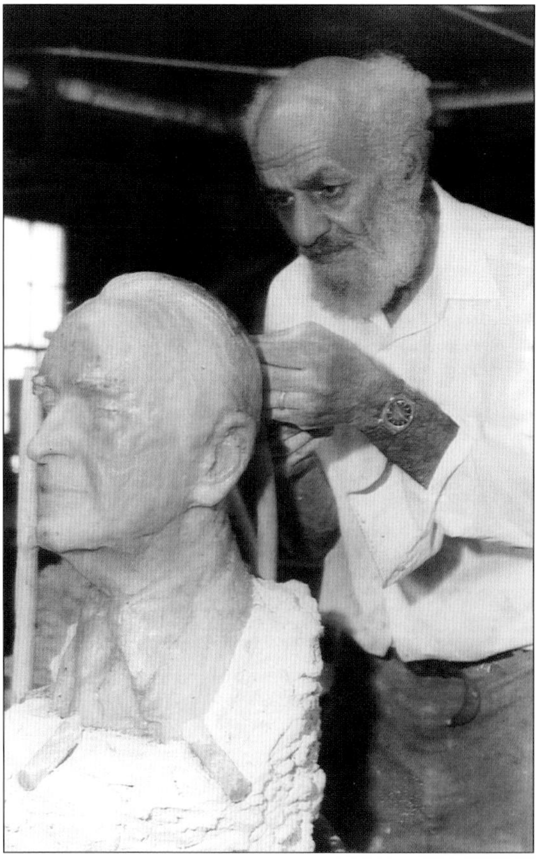

The Polish garden centered on a sculptured bust of Ignacy Jan Paderewski (above), the renowned concert pianist and composer who became Poland's first prime minister in 1919 and represented his country at the Paris Peace Conference. The Michigan City Paderewski Society, headed in the 1950s by Sophia C. Kominiarek, commissioned local sculptor Robert Wilcox (left) to create the terra-cotta sculpture. Actually, he made two busts of Paderewski. After the first was damaged, he did an acrylic piece in the 1970s. That one has also vanished, and another has recently been commissioned.

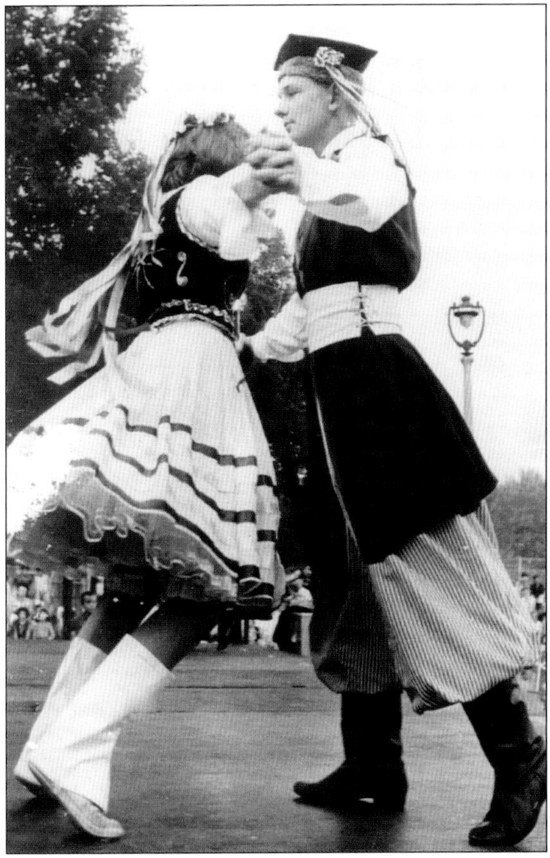

The Mazur dancers of Milwaukee performed Polish polkas (right) and the "switching day" dance (below), a special routine celebrating Dyngus Day, the Monday after Easter. Polish festivities have taken place annually in the gardens and were pictured here in 1957. (Both, courtesy of the *Milwaukee Journal*.)

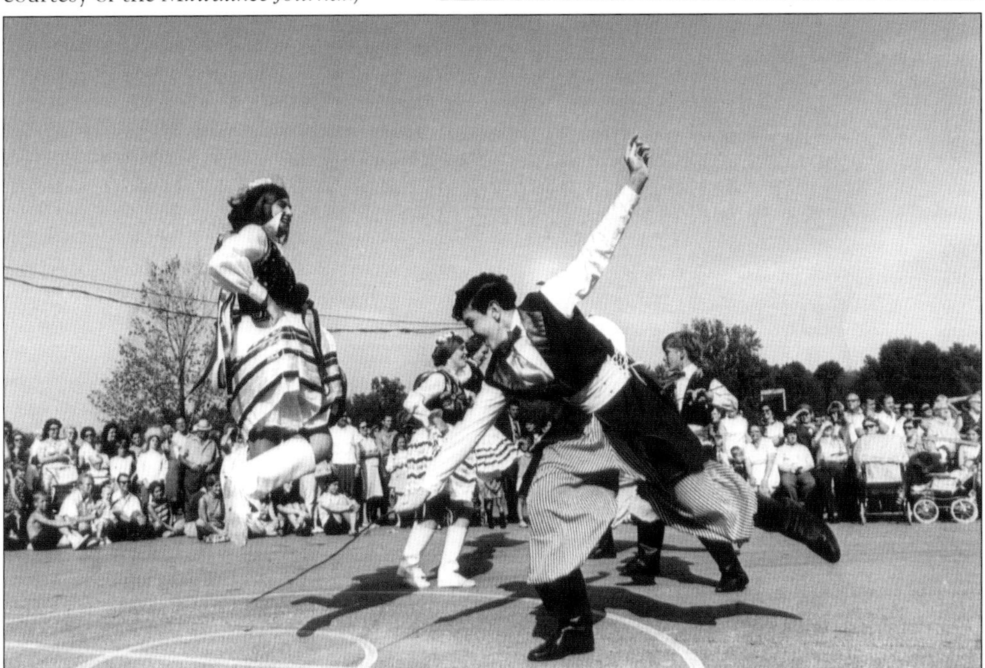

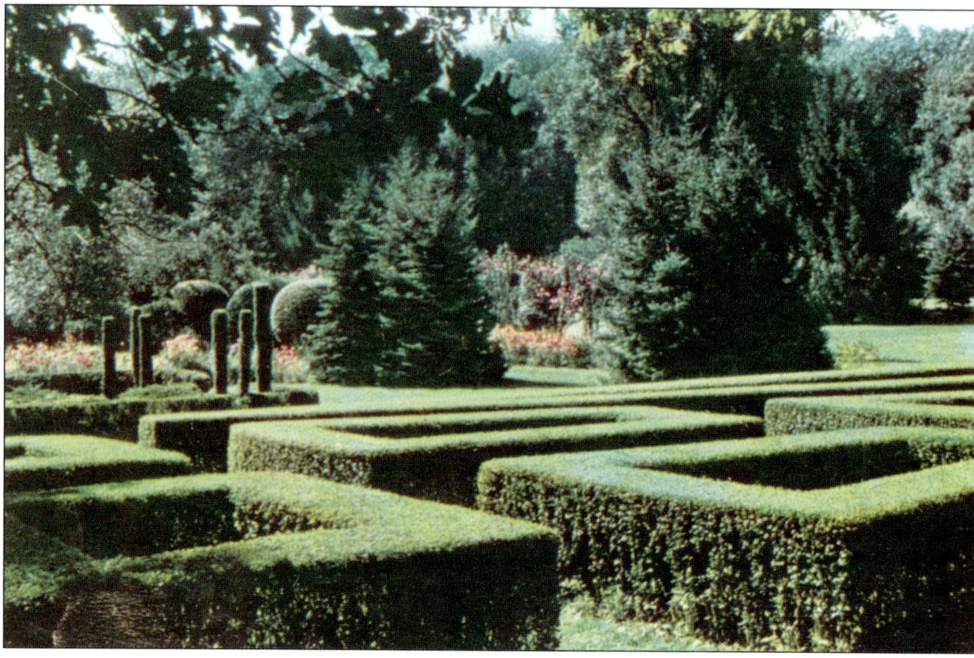

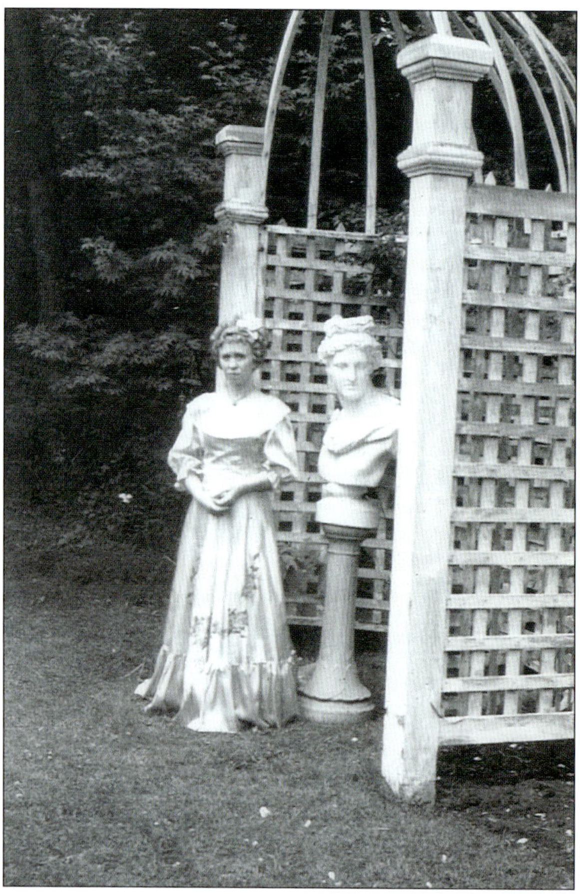

The characteristic of French gardens replicated here (above) is a maze created by neatly trimmed hedges. At left, a demure young lady is pictured beside a piece of formal statuary, sheltered by a latticed canopy in the French garden.

The Italian garden, a scenic spot, featured a reflecting pond and a cherub lifting a platter, which had previously been shown in the Chicago world's fair horticultural exhibit. Around the perimeter of the Italian garden (below) is a series of reproduction Corinthian columns referencing the Roman Forum. Italian leaders had chosen the column as a symbol of their country's participation in the fair: one ancient column, a gift to the City of Chicago, still stands in Burnham Park, the only piece remaining in the fair's original location. (Right, photograph by Marshall Newspictures of Indianapolis; courtesy of Jeanne Russ.)

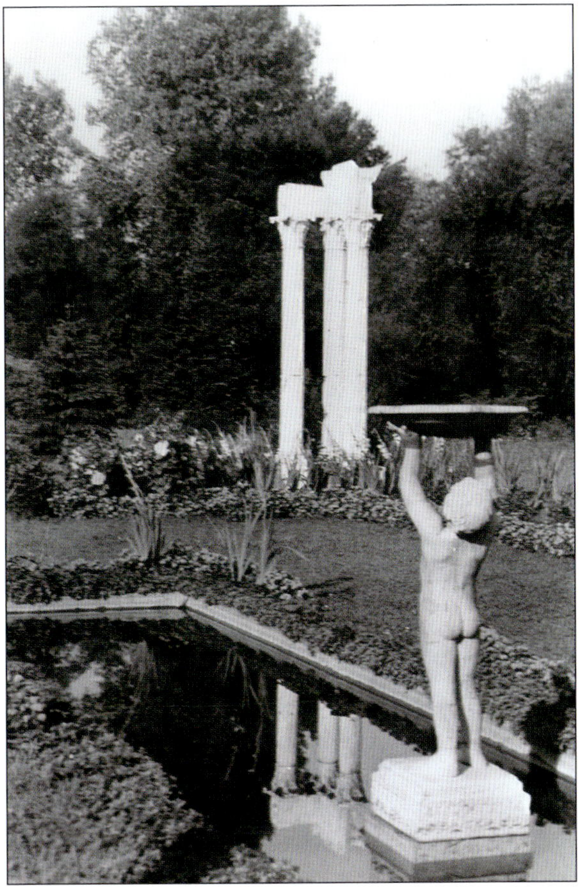

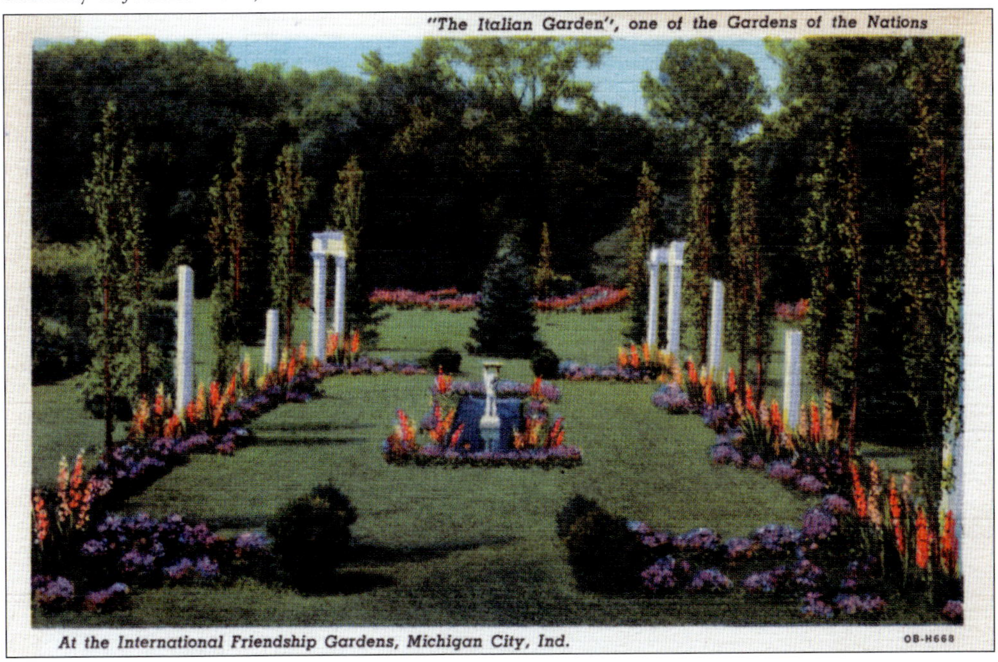

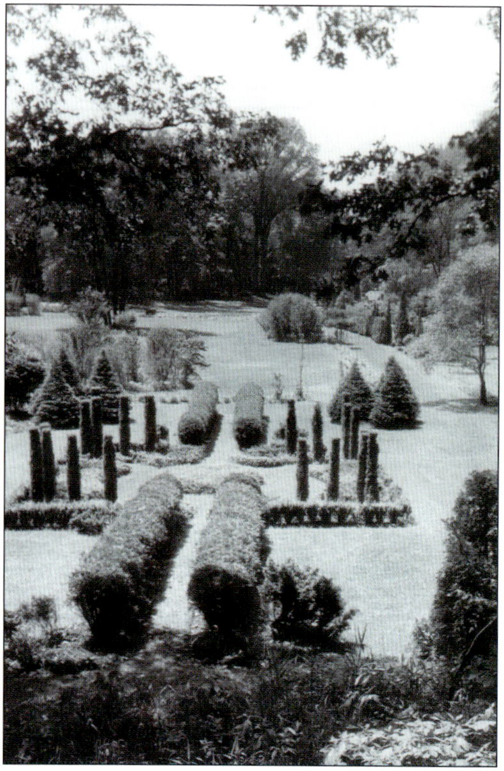

The formal Grecian garden had an arrangement of tall junipers, trimmed to resemble columns and planted in a perfect square. The plan was to have Grecian statuary placed between the trees but, according to the gardener's notes, "This will now have to wait until after the War." Evidently referring to World War II, this comment dates the photograph to the early 1940s.

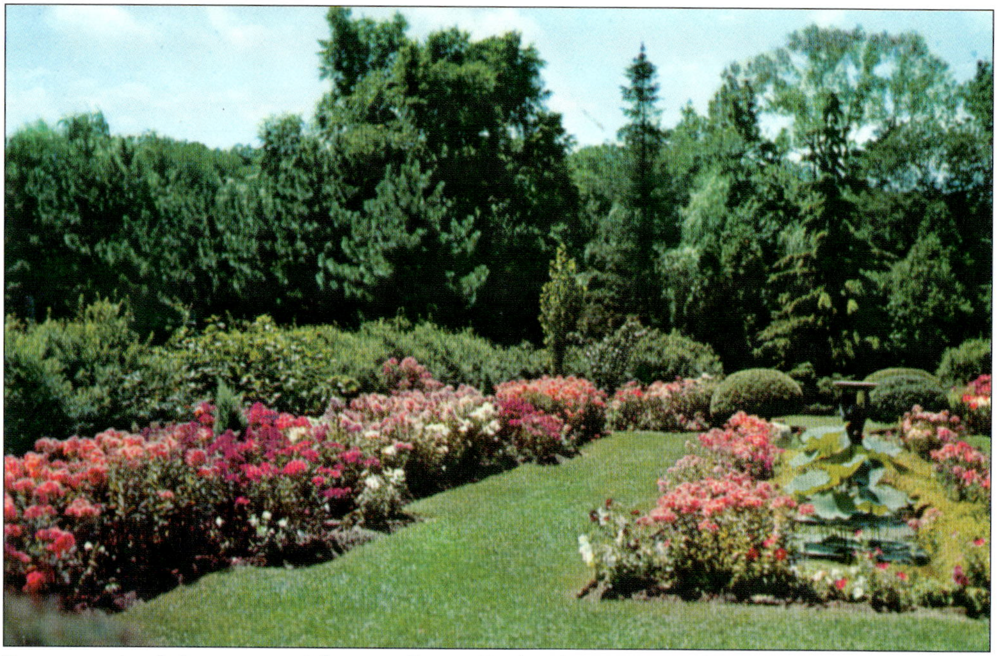

Roses remained the perennial favorite flower. By far the most popular in the Stauffers' booklet of international favorites, the rose dominated the original Persian garden as well as other ethnic gardens. In 1953, the rose took on a more commercial aspect, when Jackson and Perkins Company granted a license to International Friendship Gardens to grow roses for the purpose of testing.

Striking a statuesque pose, this unidentified model gestures toward the Grecian garden.

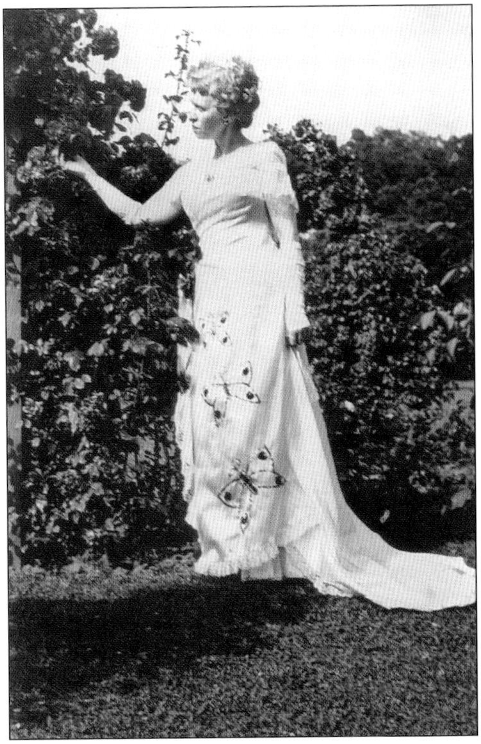

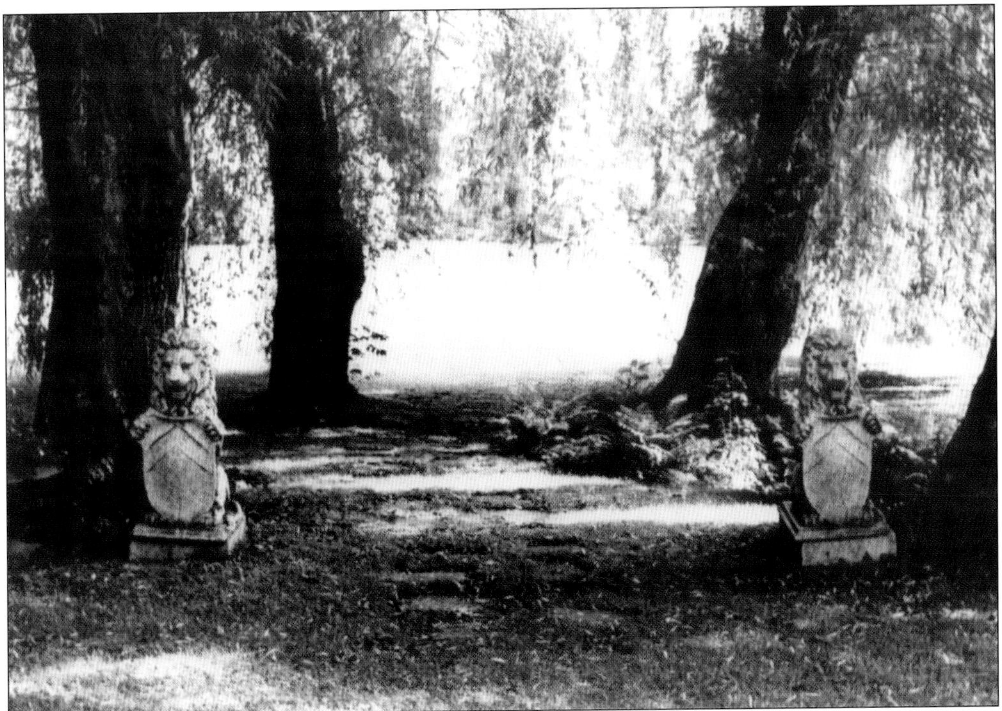

A pair of lions, often selected to symbolize royalty, guarded the entrance to the Chinese garden. When the mighty lion is holding a heraldic shield, as in this case, the affiliation with royal families is reinforced. The lions are still in the gardens, though in a different location.

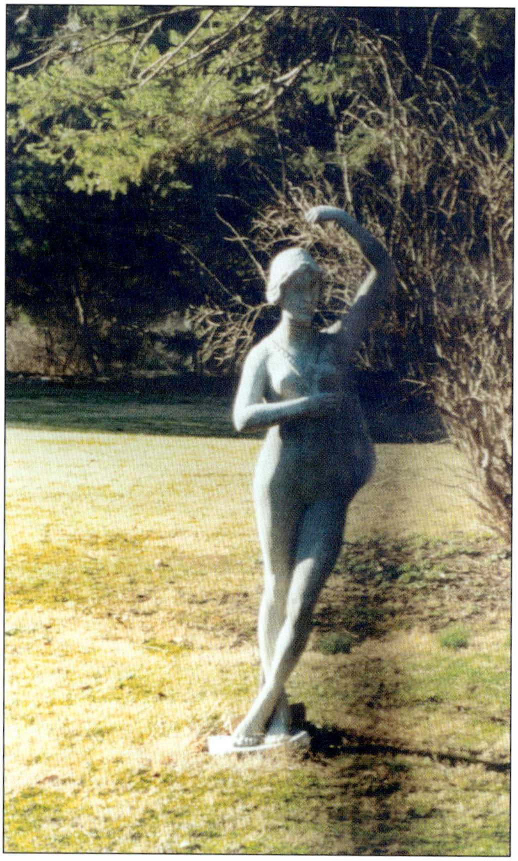

Life-sized figures attributed to Mustafa Naguib, an Egyptian sculptor who studied in Rome, Italy, and then immigrated to northwest Indiana, are placed throughout the gardens. Naguib founded a school of sculpture in Hobart, Indiana, in 1973 and then moved to Beverly Shores, Indiana, where he had a hilltop studio for several years. His plans to establish an art school at International Friendship Gardens never materialized. He moved to Glen Ellyn, Illinois, where he opened another school of sculpture in 1980. (Both, courtesy of Jeanne Houck.)

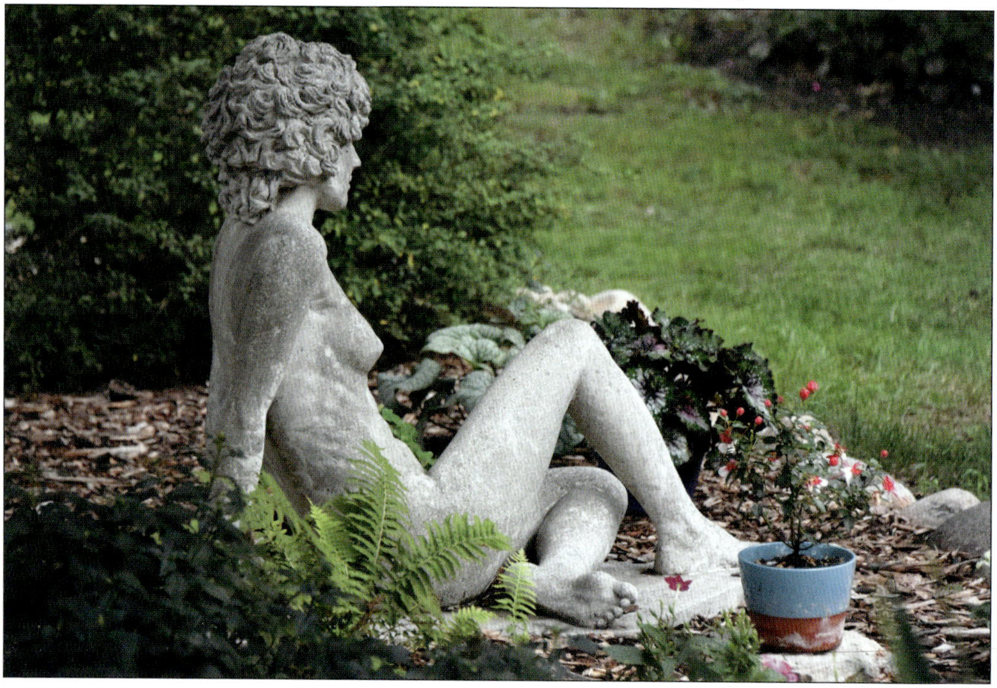

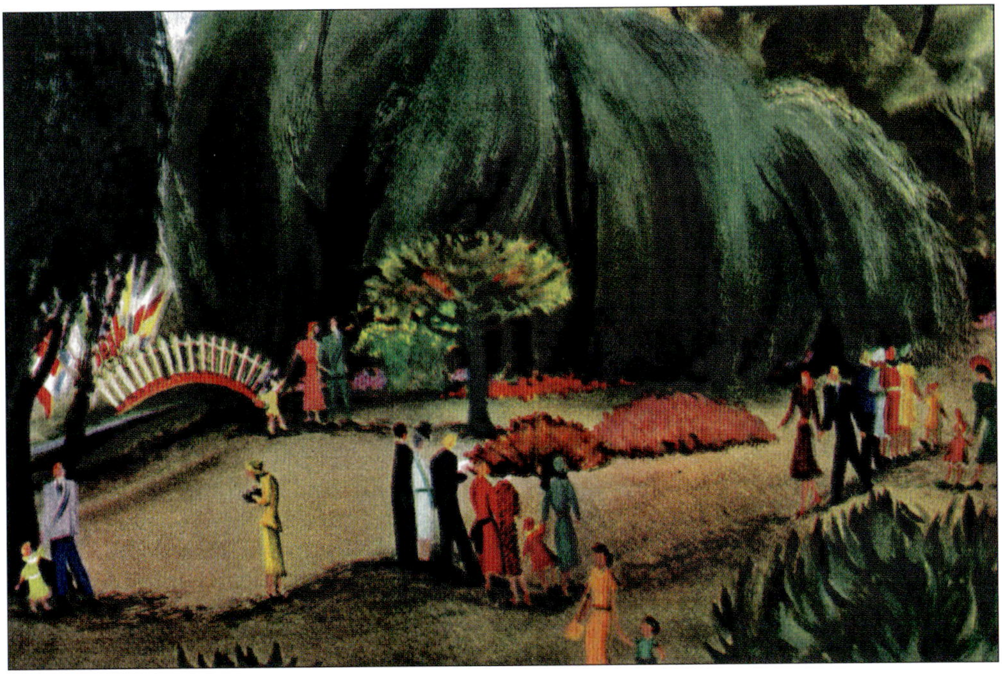

The picturesque character of International Friendship Gardens appealed to many artists. These paintings by Katherine Grace (above and below) were reproduced in 1950 in *Ford Times*, a magazine printed in Detroit, Michigan, to encourage interstate road trips. Grace exhibited her work in Delavan, Wisconsin, and won an award in the *Chicago Tribune* Art Competition.

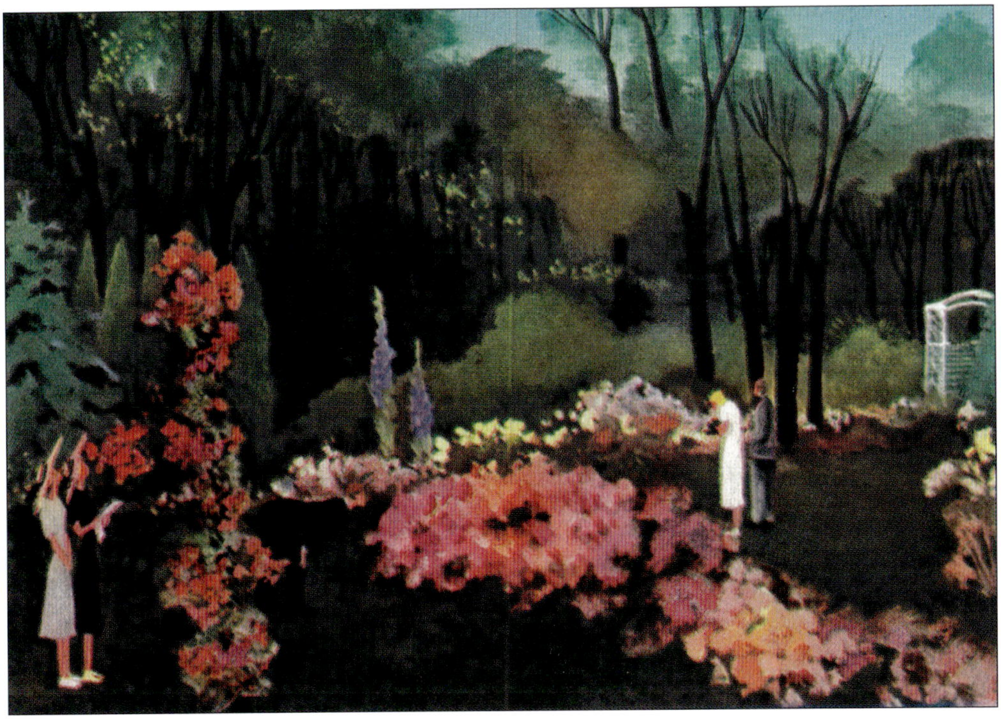

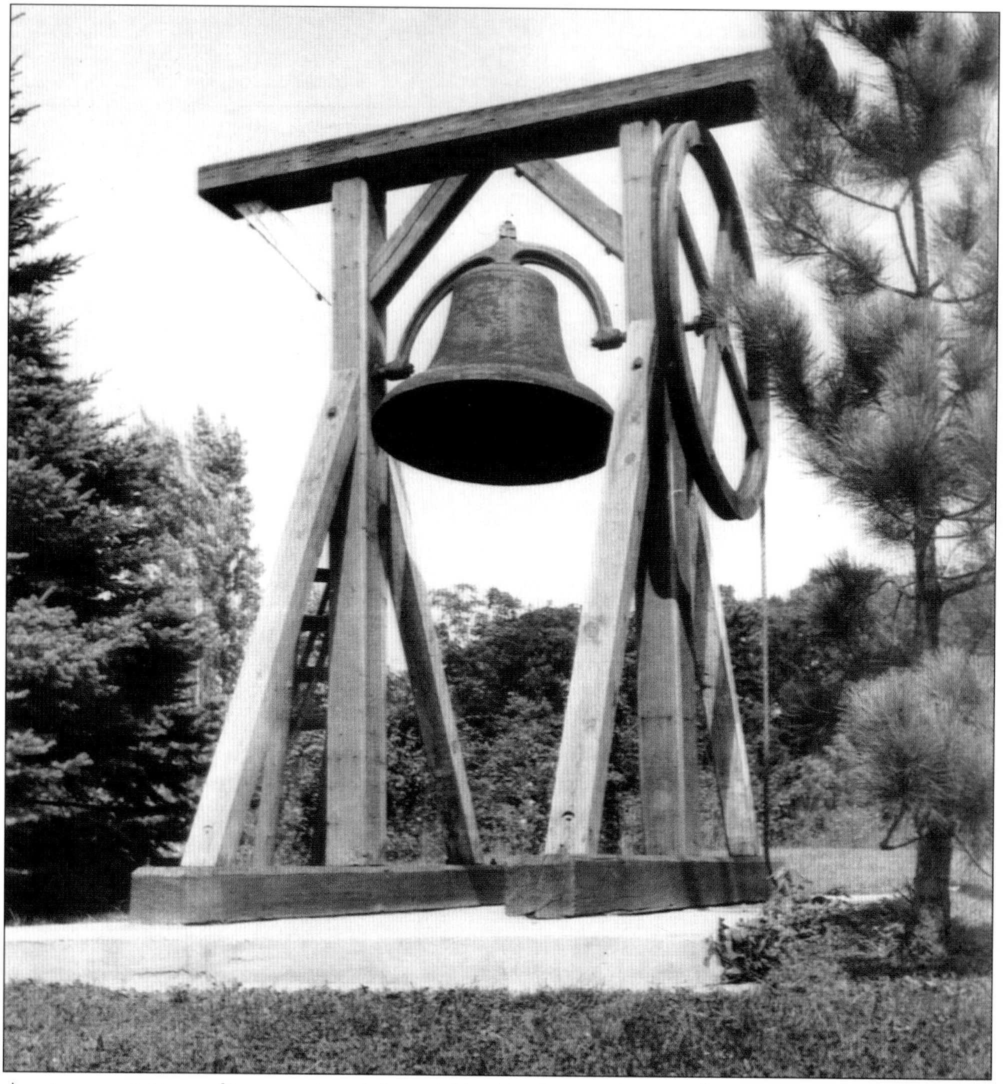

A prize acquisition of International Friendship Gardens is the Freedom Bell (also known as the Peace Bell), which marked the end of World War II. It is a smaller replica of the 10-ton bell that was permanently installed in Berlin, Germany, after its US tour of 48 states. It was designed by Walter Dorwin Teague and cast in England. In 1951, this replica was presented to the gardens by the National Committee for a Free Europe. (Photograph by Eugene Ericson; courtesy of Alex Rakowski.)

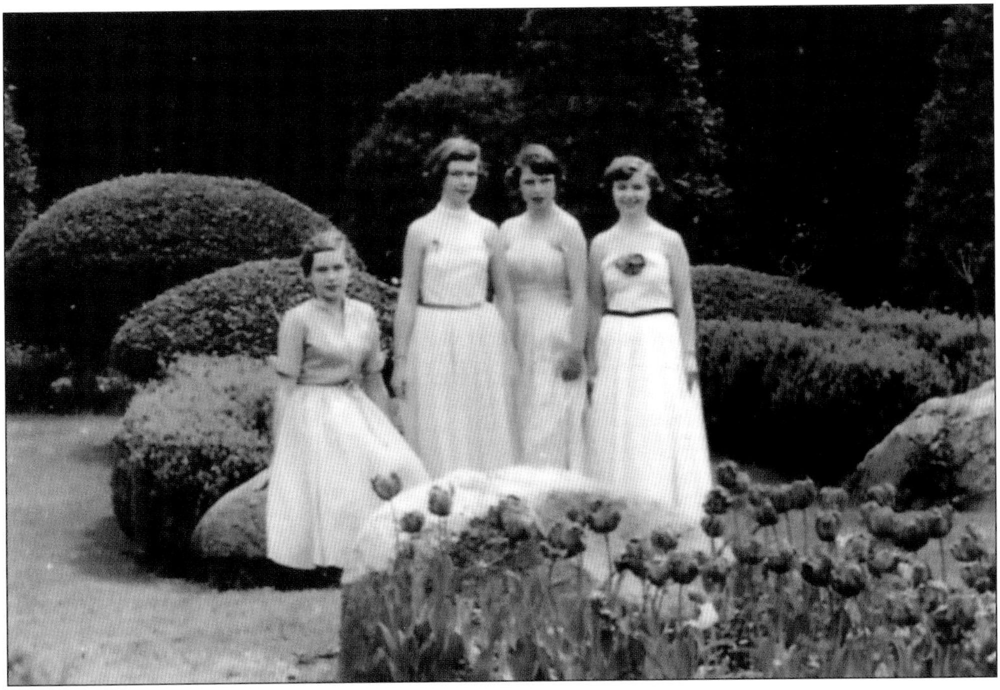

In the springtime of their lives, these young ladies are, from left to right, Deanna Gonder Caulfield, Mickey Fargher Lewis, Carol Bartholomew, and unidentified. (Courtesy of Deanna Caulfield.)

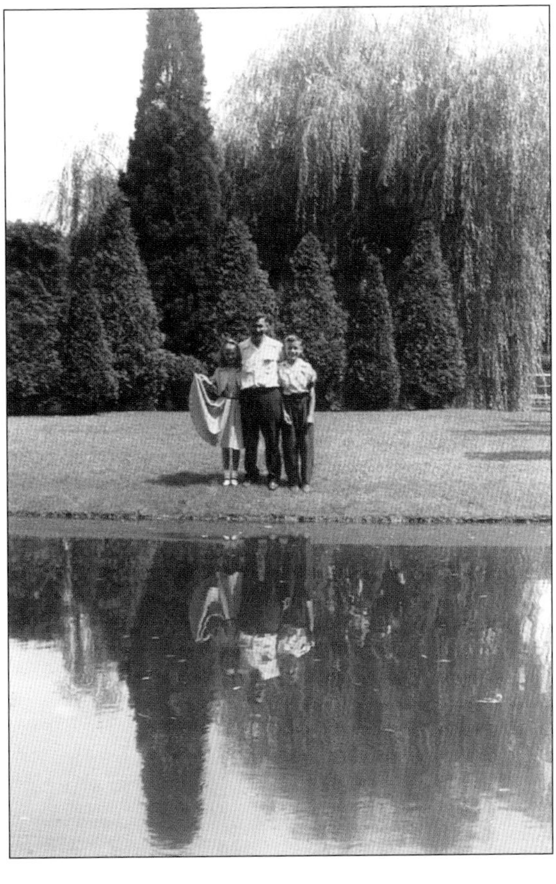

The Tylisz family, of Michigan City, is portrayed in 1948 on the shores of Lake Lucerne: From left to right are daughter Phyllis, dad Walter, and son Gene, then nine years old. (Photograph by Harriet Tylisz; courtesy of Gene Tylisz.)

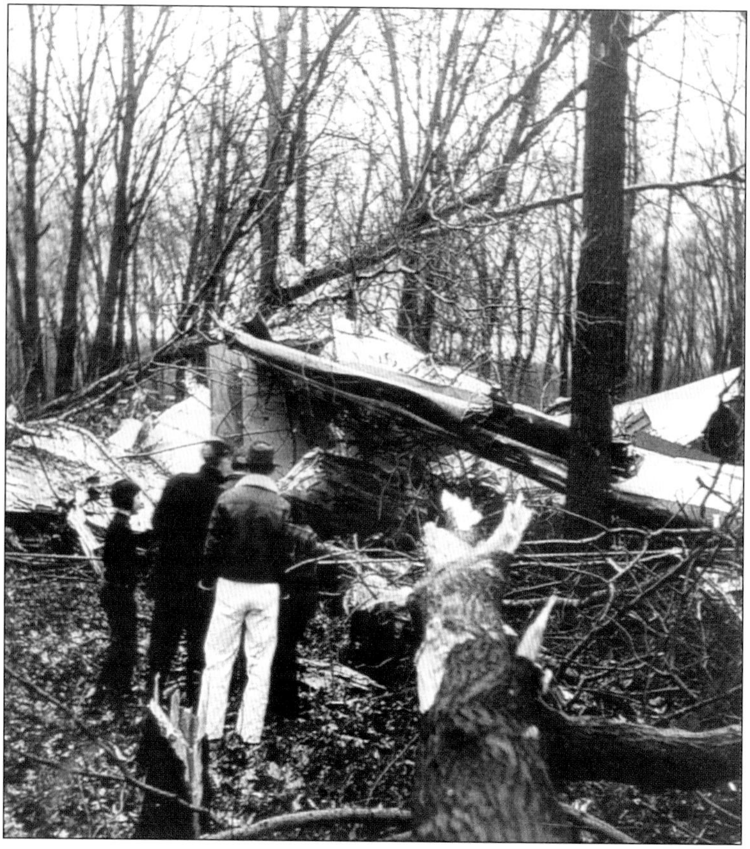

This idyllic setting became the scene of disaster: the crash of an American Airlines plane bound from Buffalo to Chicago on December 28, 1946. Both the pilot and copilot were killed, but all 18 passengers and the stewardess survived and were taken to St. Anthony Hospital for treatment. Witnesses speculated that the American Airlines DC3 had ice on its wings and was heading for the Michigan City airport, just a few miles away. (Both photographs by Walter Tylisz; courtesy of Gene Tylisz.)

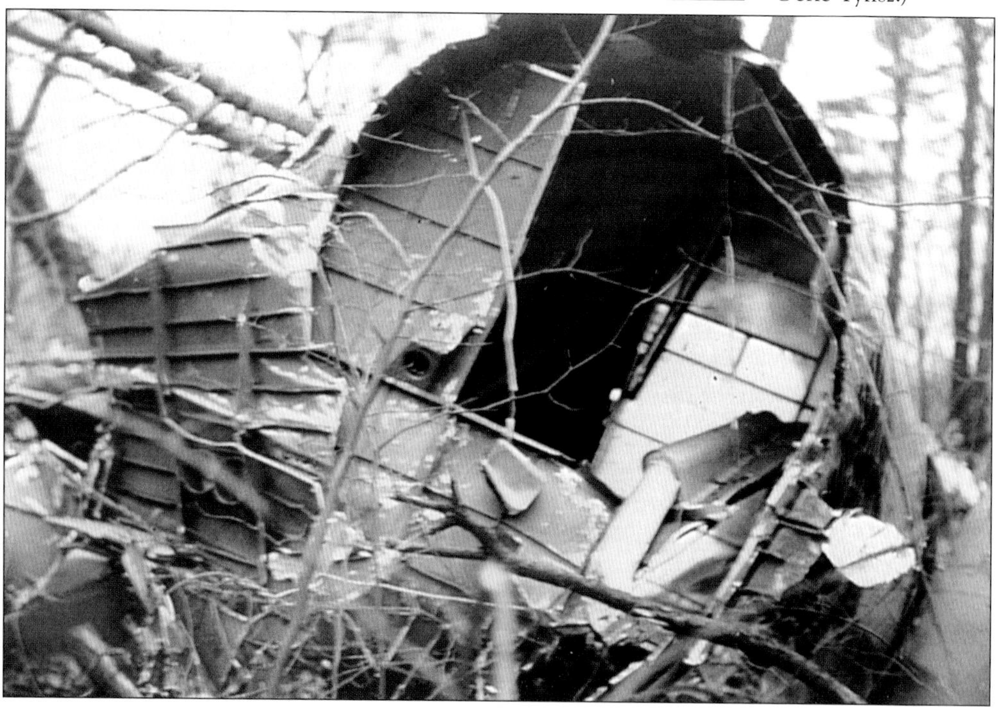

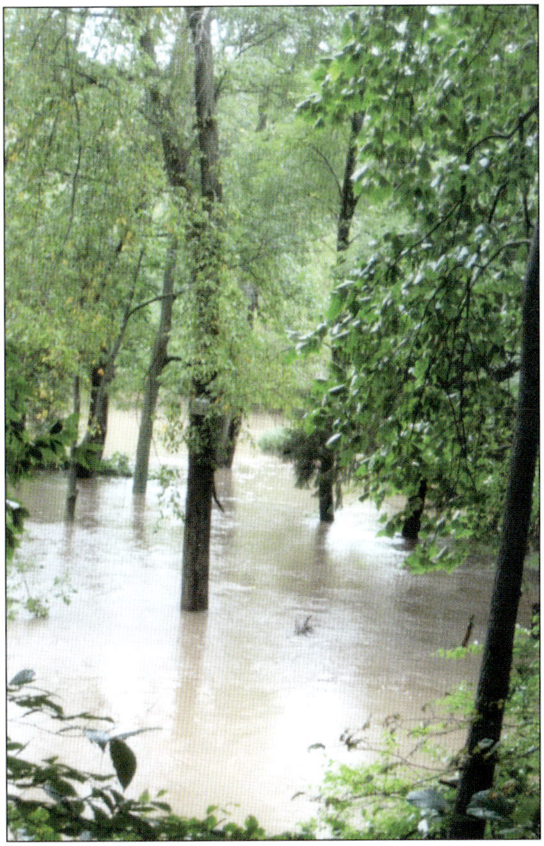

Natural disasters also occurred. In 30 years, Dr. Houck recalled, "We had five floods." Pictured here is a stagnant pool with drenched trees (right) and a rushing stream (below), threatening the safety of the guardian lion. Floodwaters wiped out the low-lying Italian gardens, including most of its Corinthian columns. Another storm destroyed the French, Greek, and Scottish gardens. (Both, courtesy of Jeanne Houck.)

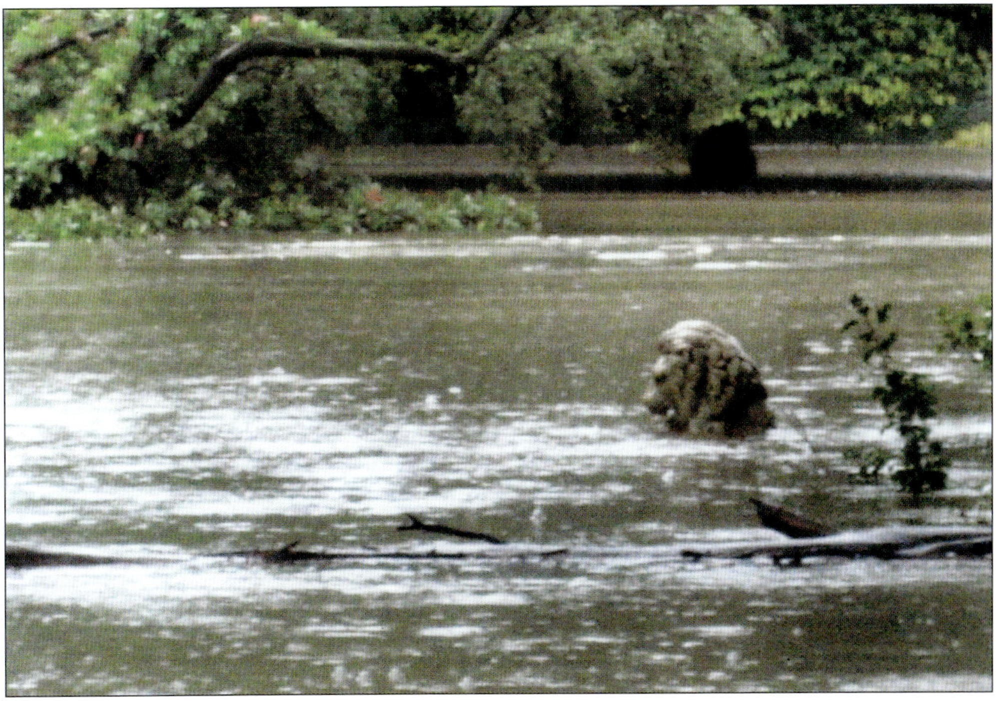

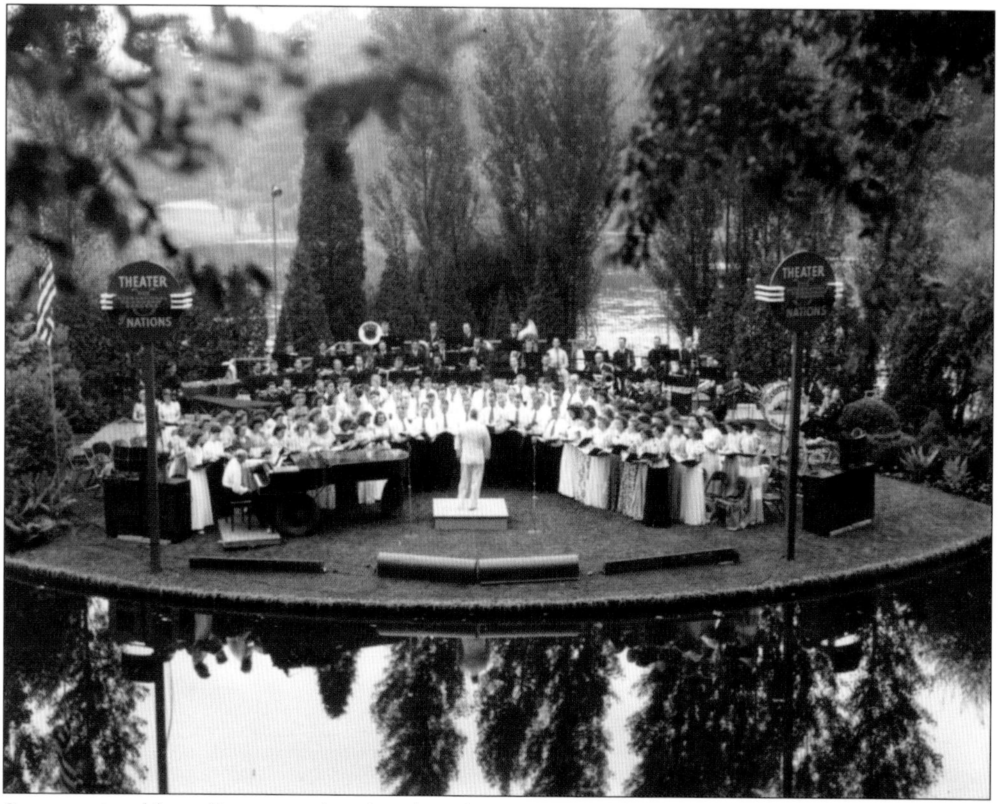

Summertime blessedly returned and with it the evening performances at Theatre of Nations on peaceful Lake Lucerne. A choral group is shown above, and opera is performed below.

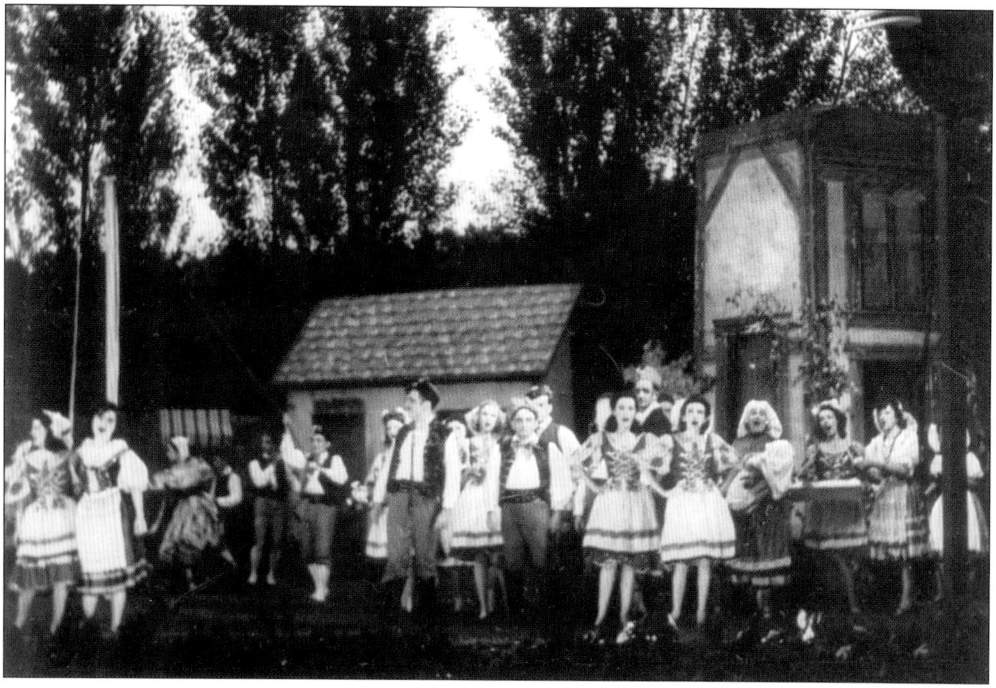

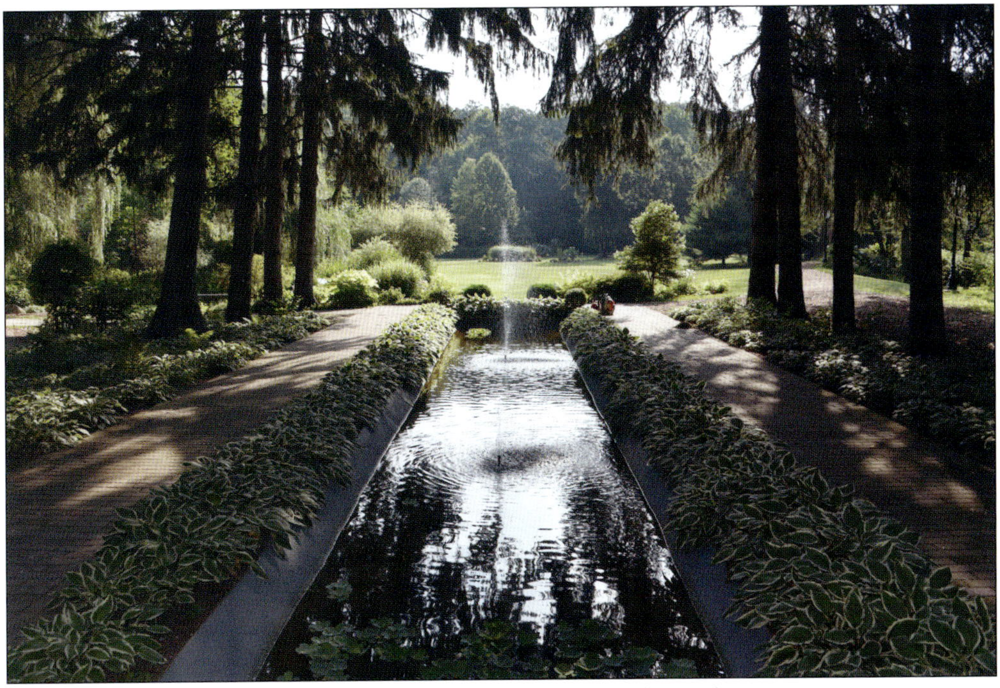

Symphony Gardens expressed the tranquility envisioned by the gardens' creators..

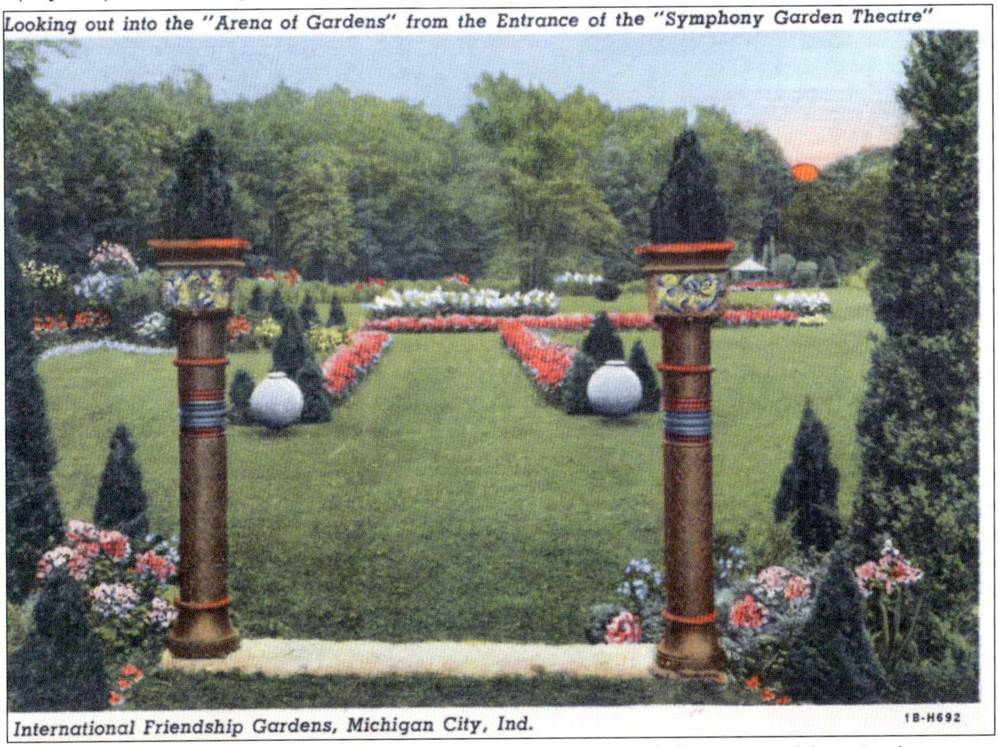

A poem by Gertrude M. Smith describing the gardens reads, "These Friendship Gardens are a dream / Of what should be our earthly theme / A harmony of light and sound / And beauty from the world around / A monument to all that's good / In International brotherhood."

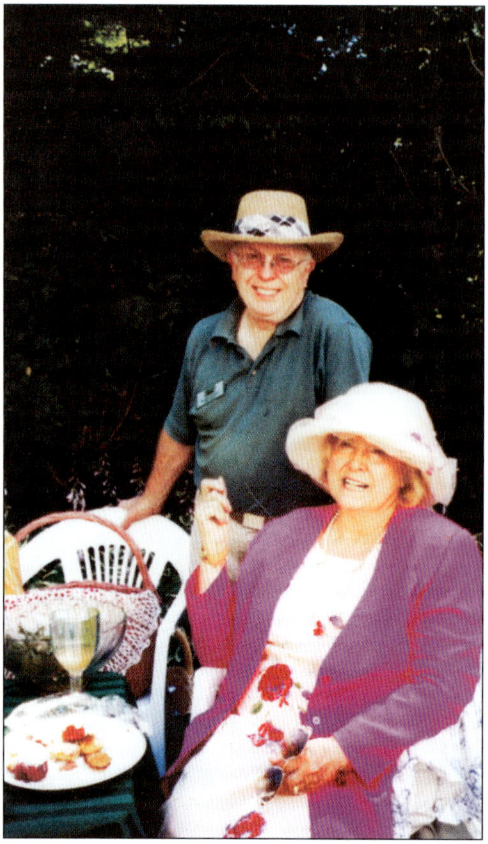

Dr. Richard Houck and his wife, Jeanne, became strong supporters of International Friendship Gardens soon after their arrival in Michigan City, Indiana. The couple had met at Indiana State Mental Hospital in Indianapolis, where he was externing and she was working as a psychiatric nurse. They have both worked as volunteers at International Friendship Gardens for over 40 years. (Courtesy of Jeanne Houck.)

Clarence Stauffer, who remained active well into his 90s, took over as director of the gardens in 1964 and remained in that capacity until his death in 1986. Clarence and his wife, Frances, lived in a small service building on the grounds. He is pictured here in the canopied structure built to shelter special events. (Courtesy of Jeanne Houck.)

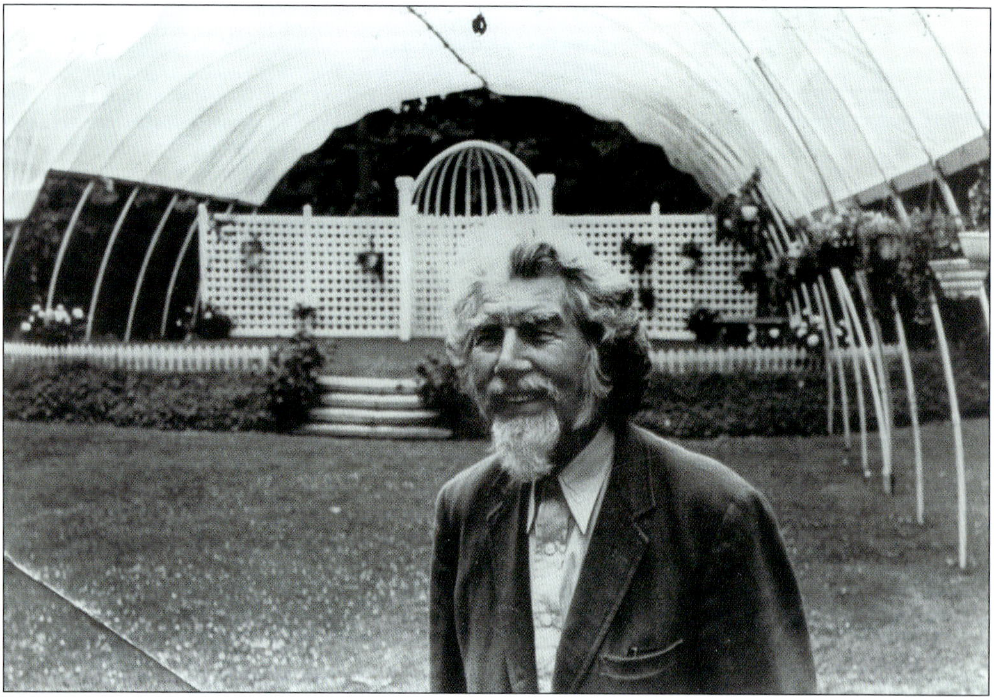

Four

Jeanne Houck and Clarence Stauffer

In 1970, Dr. Richard Houck moved back to his hometown of Michigan City, Indiana, to open a private ophthalmology practice. For 12 years, he and his wife, Jeanne, had lived in 10 different states, the District of Columbia, and Paris, France, while he served on active duty with the US Air Force. Newly retired as a full colonel, he and Jeanne continue to live in their historic home in Long Beach, Indiana.

Knowing of his wife's love for gardening, Dr. Houck suggested she check out the situation at International Friendship Gardens, which he remembered from his childhood. Jeanne's intervention could not have come at a better time. The youngest of the Stauffer brothers, Clarence, then 84, was struggling to maintain the gardens on his own, with dwindling resources and an ailing wife.

Jeanne brought youthful energy and enthusiasm to the project. She contacted local organizations to help rejuvenate what still remained of the ethnic gardens. She hounded local nurseries and charitable groups for contributions. She organized afternoon tea parties and evening concerts to restore the genteel, cultural atmosphere of the place. She initiated new plantings, most importantly the heart-shaped garden of Peace roses, as a tribute to American veterans. She publicized peak blooming seasons from April through October.

Clarence continued giving tours to the groups of people who visited International Friendship Gardens. He welcomed the assistance of Jeanne, who not only worked tirelessly in the gardens, but befriended his wife, Frances, and enriched her life by introducing her to "talking books." In 1973, Clarence designated Jeanne as "The Lady of the Gardens," a title confirmed by the board in 1999, when she was named executive director emeritus.

Frances Willard Stauffer died in 1975 at the age of 76, and Clarence survived her by another 11 years. He died in 1986 at age 100.

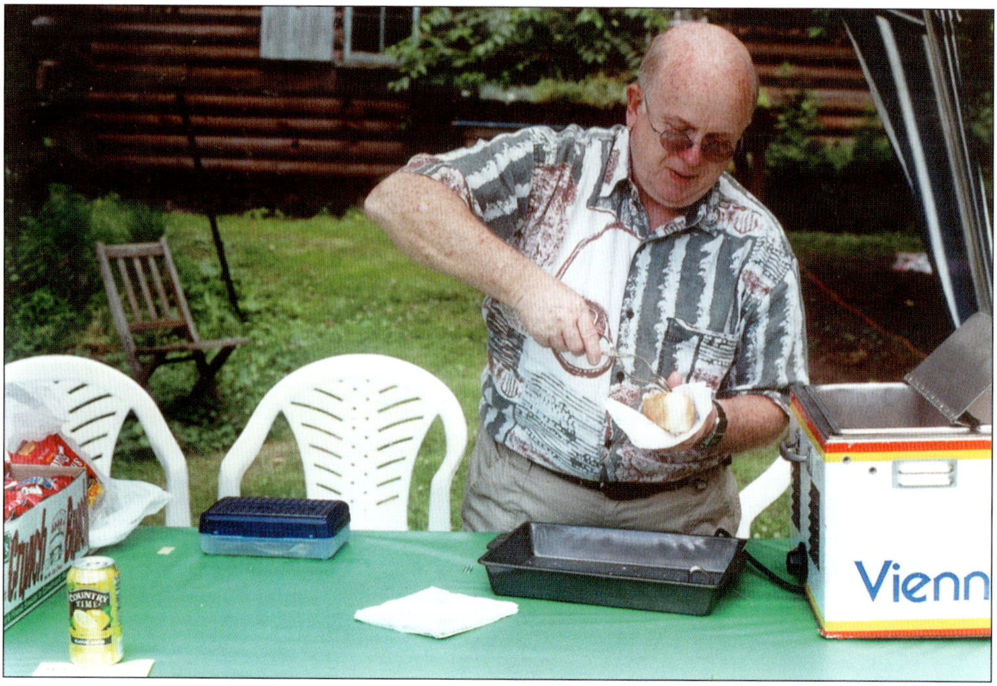

An outgoing person, Dr. Houck pitched in at special events wherever he was needed. For 20 years, he has served on the gardens' board of directors, currently as emeritus, alternating between positions of president and treasurer. (Courtesy of Jeanne Houck.)

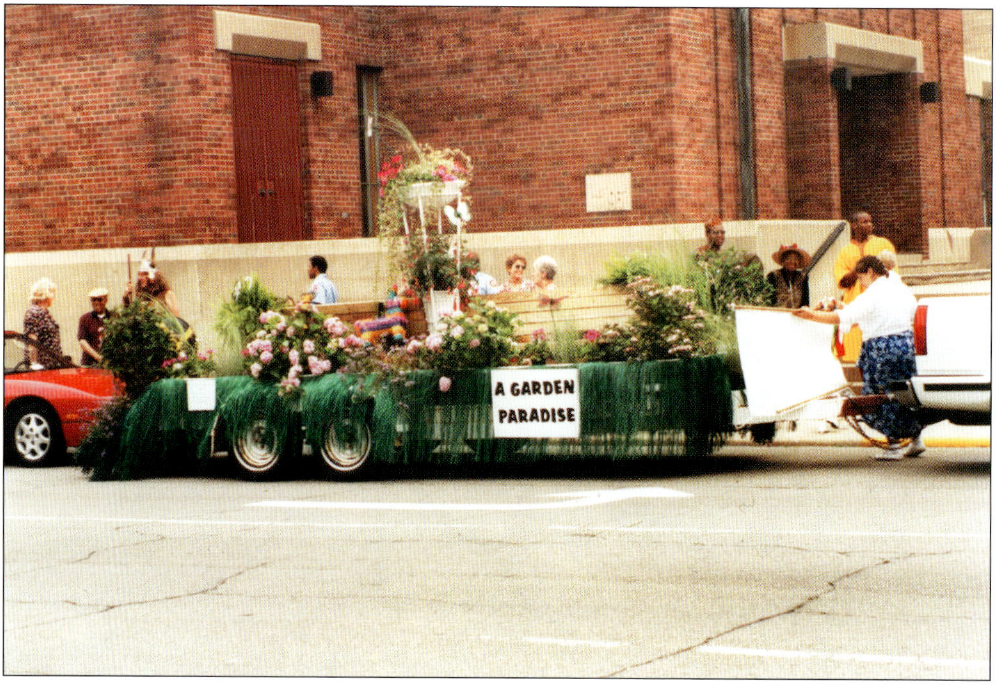

Participation in downtown Michigan City parades helped maintain a higher profile within the community, while also attracting younger volunteers to work at International Friendship Gardens. In 2000, Jeanne Houck was grand marshal of the parade. (Courtesy of Jeanne Houck.)

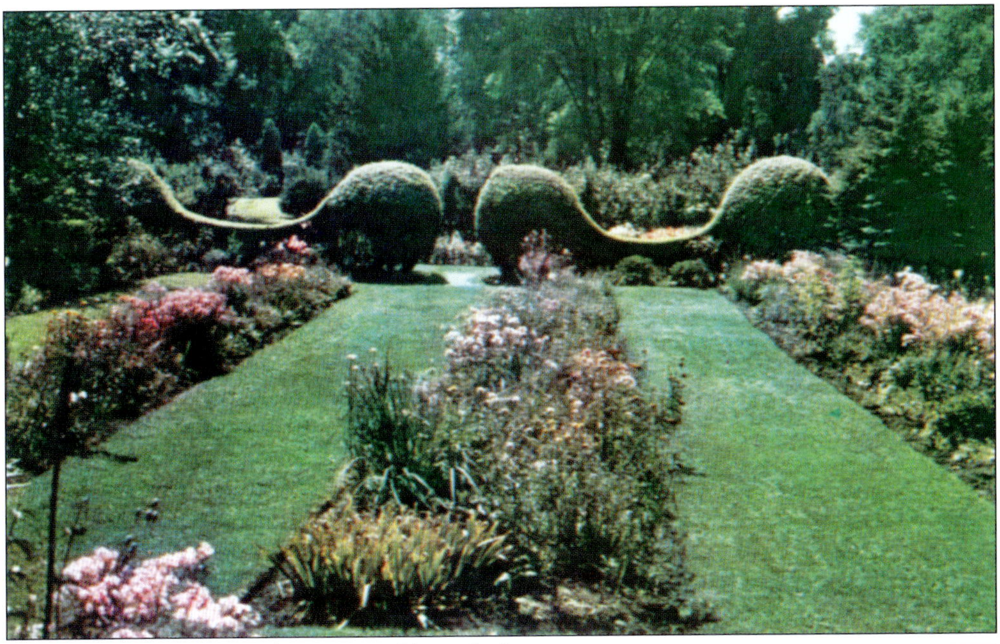

The original Scottish garden, pictured here, had a unique Scotch cradle fence. After being ruined in a flood, the Scottish garden was later restored.

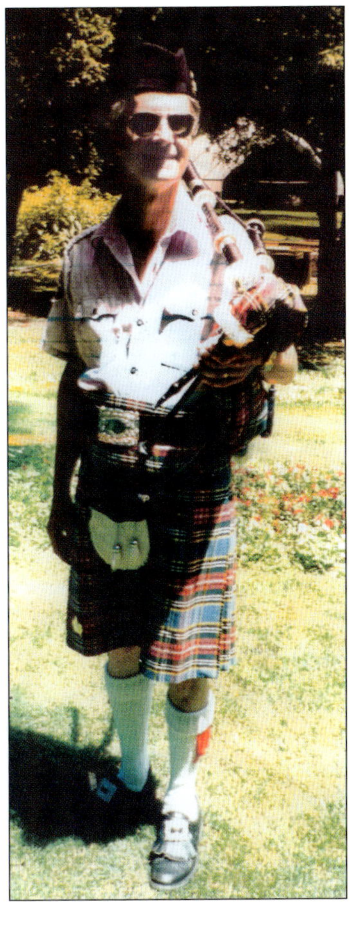

In keeping with the spirit of the Scottish garden, a bagpiper dressed in kilts contributed music to special events.

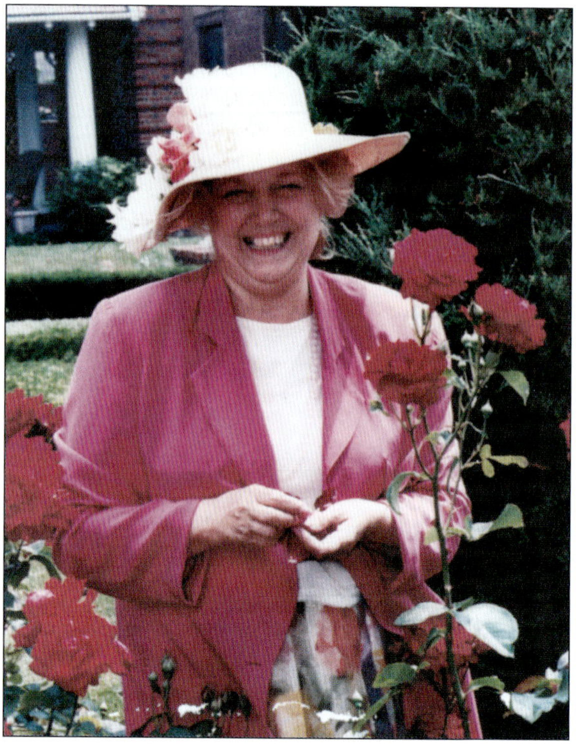

Jeanne Houck received many honors for her 47 years of service to International Friendship Gardens. In 1992, the Daughters of the American Revolution presented her with its National Conservation Award. In 1993 and again in 1999, she was named Sagamore of the Wabash, Indiana's highest honor. Leo Sharp developed a new hybrid daylily in 1999 and named it in her honor, "Lady of the Gardens." A native Kentuckian, Jeanne honed her gardening skills in various places—from Lady Bird Johnson's Bluebonnet Brigade in Texas to the gardens of Versailles, outside Paris. (Courtesy of Jeanne Houck.)

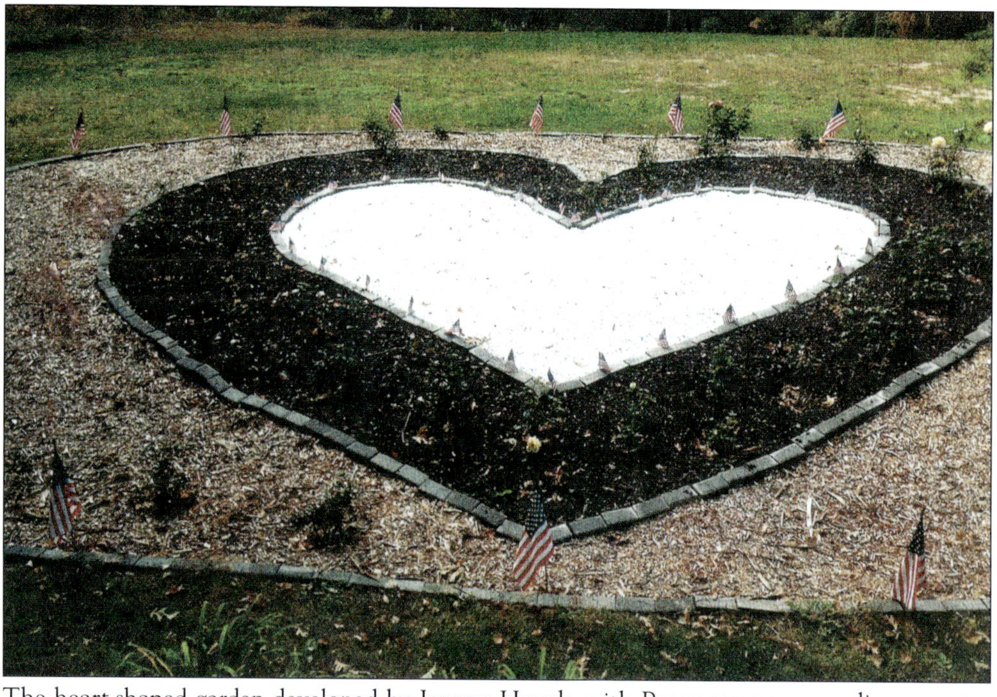

The heart-shaped garden developed by Jeanne Houck, with Peace roses surrounding a center of marble chips, was designed to welcome home American veterans. The marble chips were intended to honor those servicemen and women who gave their lives for their country. (Courtesy of Jeanne Houck.)

The picturesque atmosphere of International Friendship Gardens inspired many works of art over the years. One collage, by La Porte artist Mary Borg, depicted a fashionable Victorian lady in a red lace gown. It was dedicated to Jeanne Houck. (Courtesy of Jeanne Houck.)

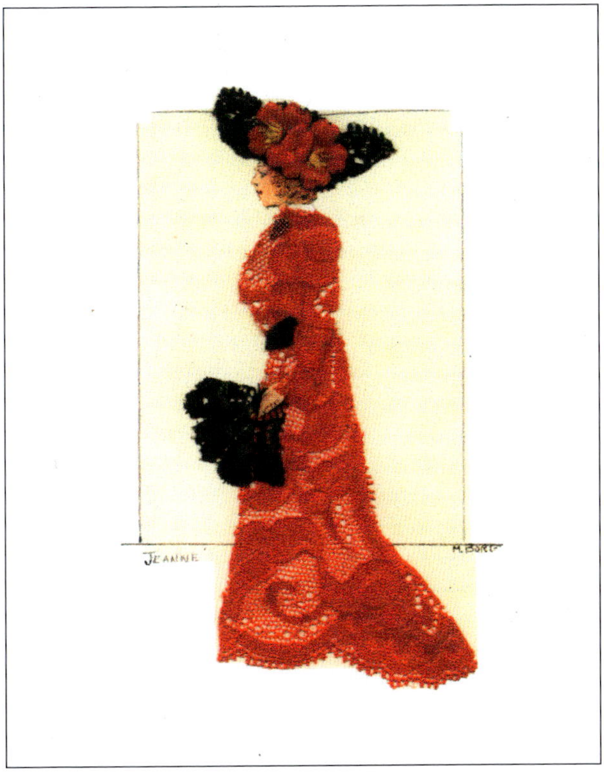

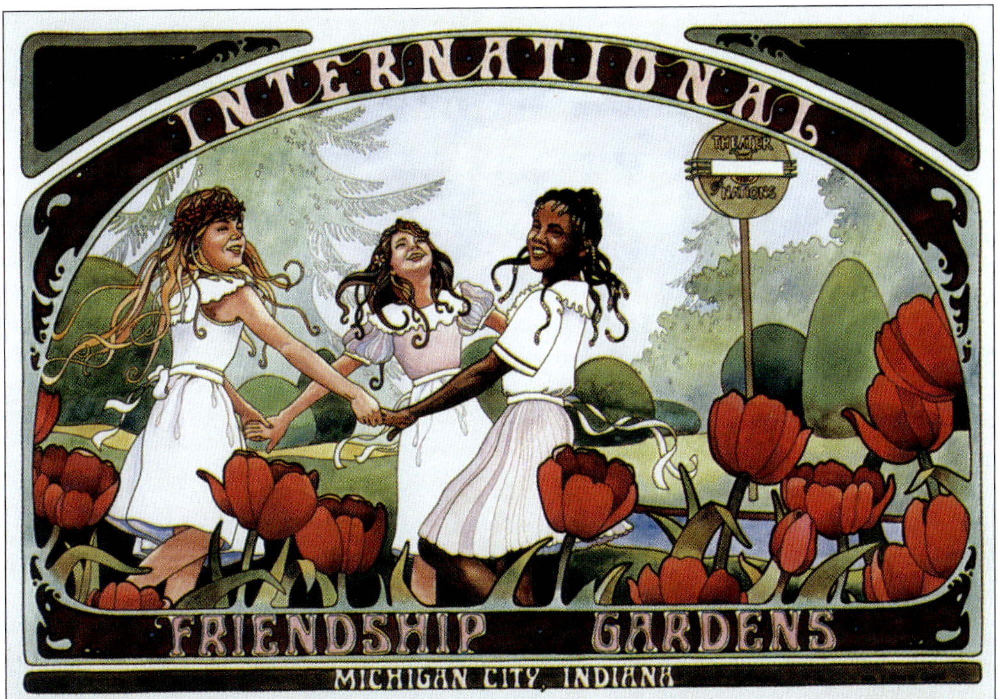

Michigan City artist Neil Kienitz created an Art Deco–style drawing that captured the multicultural character of the gardens, with three maidens dancing among the tulips that bloom in the spring.

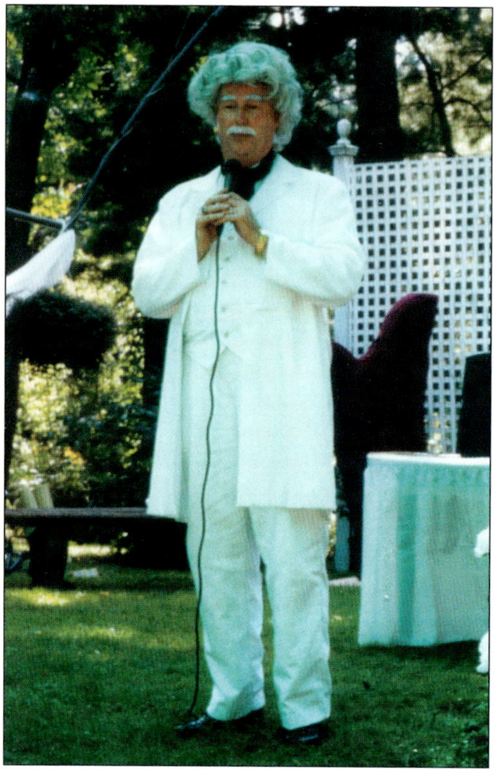

A variety of entertainers, such as a modern-day "Mark Twain," reflected the literary and historical inclinations of the gardens' founders. Emily Dickinson and other famous authors were also honored at tea parties in the gardens. (Courtesy of Jeanne Houck.)

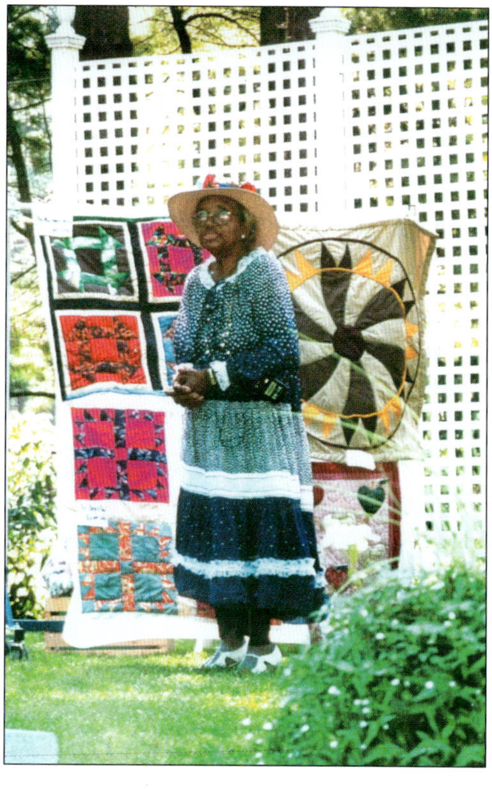

Mamie Davis, a popular performer, amazed her audience with true family stories based on the quilt collection that traced her ancestry from Southern slave plantations to present-day Michigan City, Indiana. Her quilts became part of a national touring exhibit. (Courtesy of Jeanne Houck.)

A bit of humor enlivened the ethnic festivals, as Jeanne Houck extended greetings to her "Viking" friend, Glen Malwyn. Ethnic events were sponsored annually by Lithuanian and Polish organizations and, in recent years, by groups of Native Americans. (Courtesy of Jeanne Houck.)

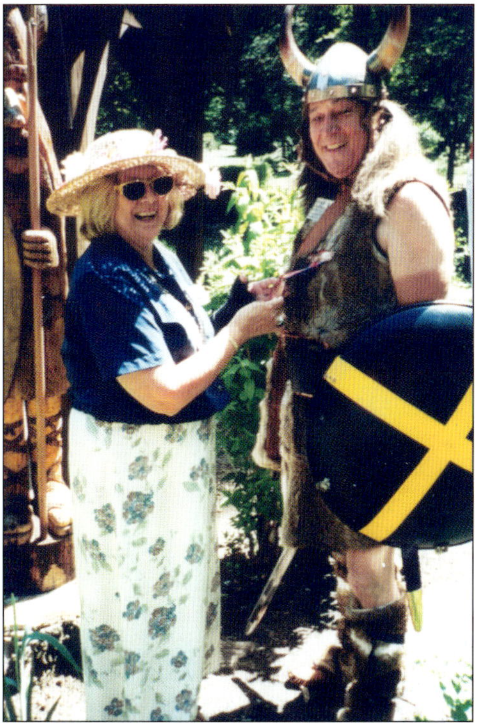

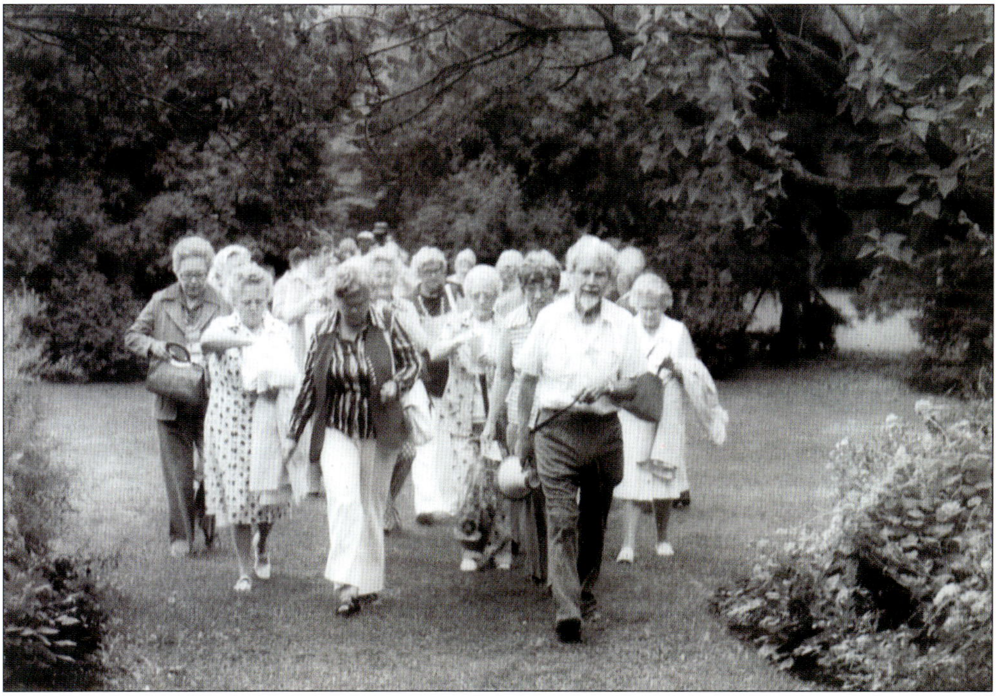

Clarence Stauffer, described by Jeanne as "very popular with the ladies," enjoyed leading tours for garden clubs and other day-trippers. Schoolchildren were intrigued by his white hair and beard. Drawing on their own experience with Kentucky Fried Chicken, they used to call out, "Look! There's Colonel Sanders!" (Courtesy of Friendship Botanic Gardens.)

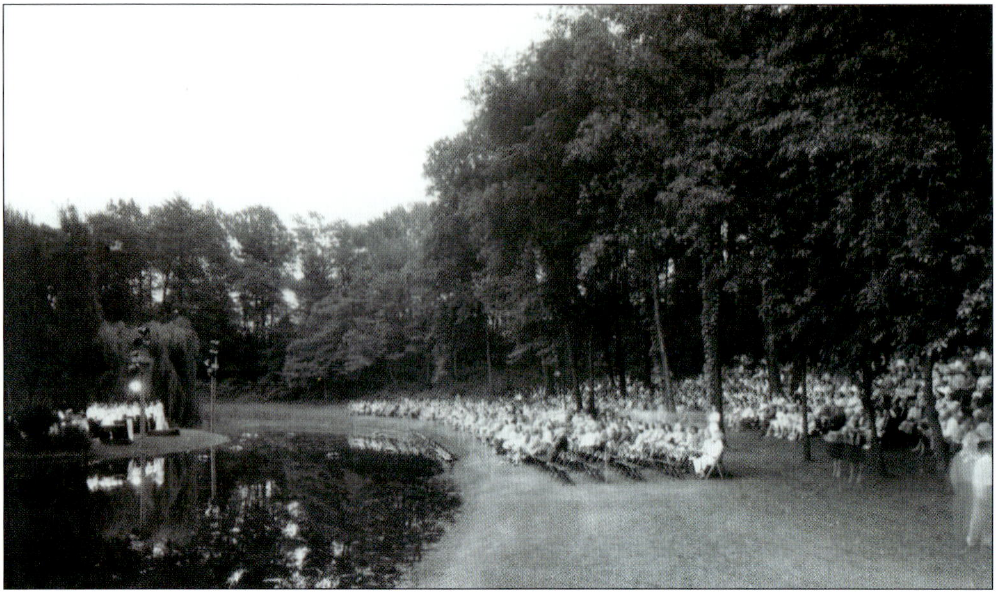

A crowd has gathered in early evening, c. 1950, and is seated on the hillside facing the Theatre of Nations on Lake Lucerne. The island stage allows for such realistic happenings as Giuletta's arrival by boat in *Tales of Hoffman*. The performance will be reflected in the tranquil waters. (Courtesy of Alex Rakowski.)

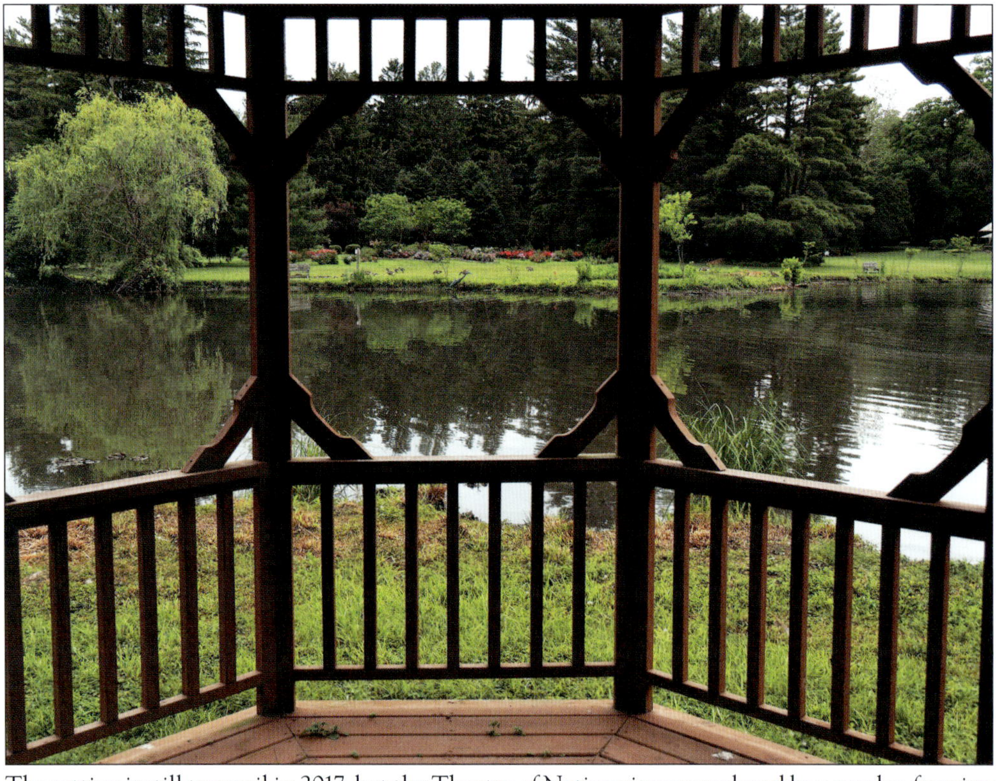

The setting is still tranquil in 2017, but the Theatre of Nations is now replaced by a gazebo, framing the view of the shoreline, and lakeside performances are replaced by contemplation.

Five

Then and Now

International Friendship Gardens endured some tough times as the 20th century wore on, such as storms, floods, and vandalism. In 2002, fire destroyed the Alaska cabin and a service building, taking with it many irreplaceable historic documents. The gardens experienced a gradual decline and loss of community support.

For 50 years, this had been a Stauffer family undertaking, but it was a family with no heirs. By the time Jeanne Houck entered the scene, the grounds were already in a sad state of repair. Jeanne retired in 1999—but not until she had established the gardens' new status as a 501(c)(3) not-for-profit corporation, assuring their eligibility for tax-deductible donations. A board of directors was to be placed in control.

Although some of the ethnic gardens still exist, others have disappeared; those of Canada, Australia, Switzerland, Greece, and Turkey are gone. The Holland garden, with its 200,000 tulips, was wiped out by floods. What is left of the English garden is a teahouse. The Italian and French plantings are reduced in size and scope. In the Chinese garden, the half-moon bridge remains, and a life-sized statue of a Chinese warrior stands guard. The stone lions have both survived and have been relocated. There is a fabulous rose garden once again, but it is no longer referred to as Persian.

Strong local support keeps the Polish, Norwegian, and Lithuanian gardens going, and the Romanian garden has been restored. A new Irish garden has a Celtic cross, and an African garden is in the planning stages. But overall, the emphasis is shifting toward more contemporary types of programs.

Dig the Dunes, a popular Facebook page, sponsored a 1920s-style Jay Gatsby party. Musical evenings now include a variety of contemporary groups. Holiday events for children, such as the Easter egg hunt and Haunted Trails for Halloween, are well-attended and have led to more children's programming. The scenic beauty of the gardens continues to provide, as always, an elegant background for weddings. For many years, a lone groundskeeper, Harold Ruggles, struggled to keep the lawns mowed; today, there are three full-time groundskeepers and Friendship Botanic Gardens are well-tended and inviting.

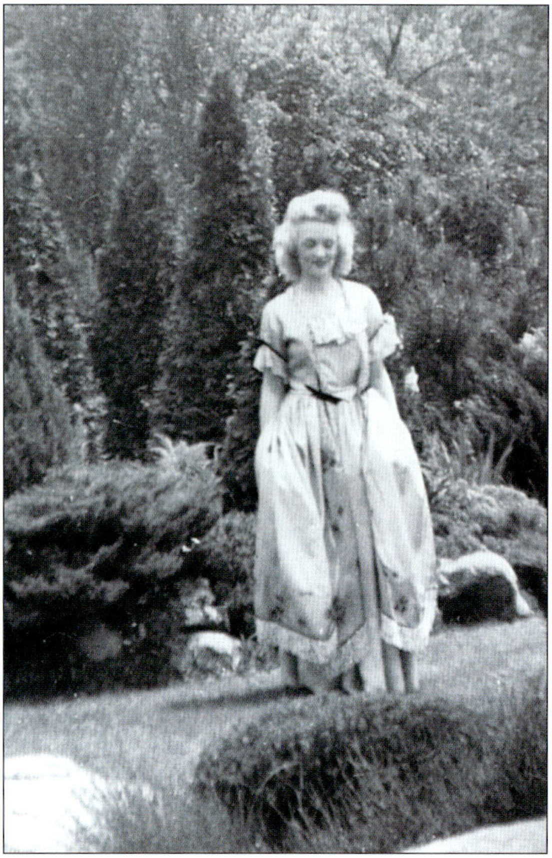

Eighty years ago, the English garden was envisioned by Clarence Stauffer as "a vast green sward that contributes so much to the beauty of the Britishers great estates." Its Michigan City counterpart was to have winding paths, huge boulders, and scenic focal points, such as the one pictured at left. Today, all that remains is a charming little teahouse (below). (Below, photograph by Phil Eichas.)

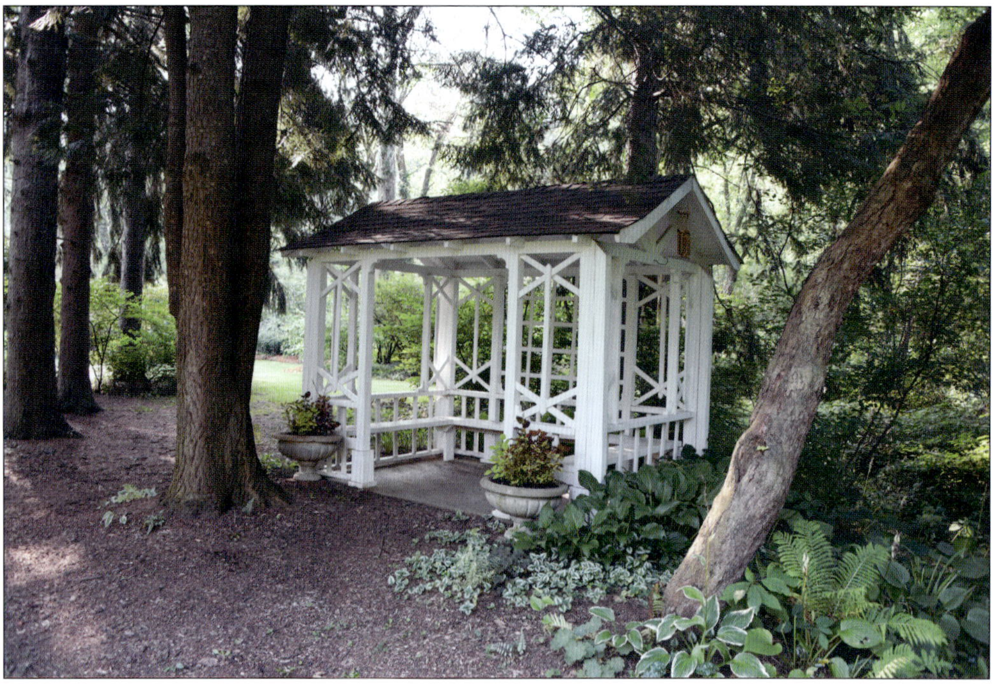

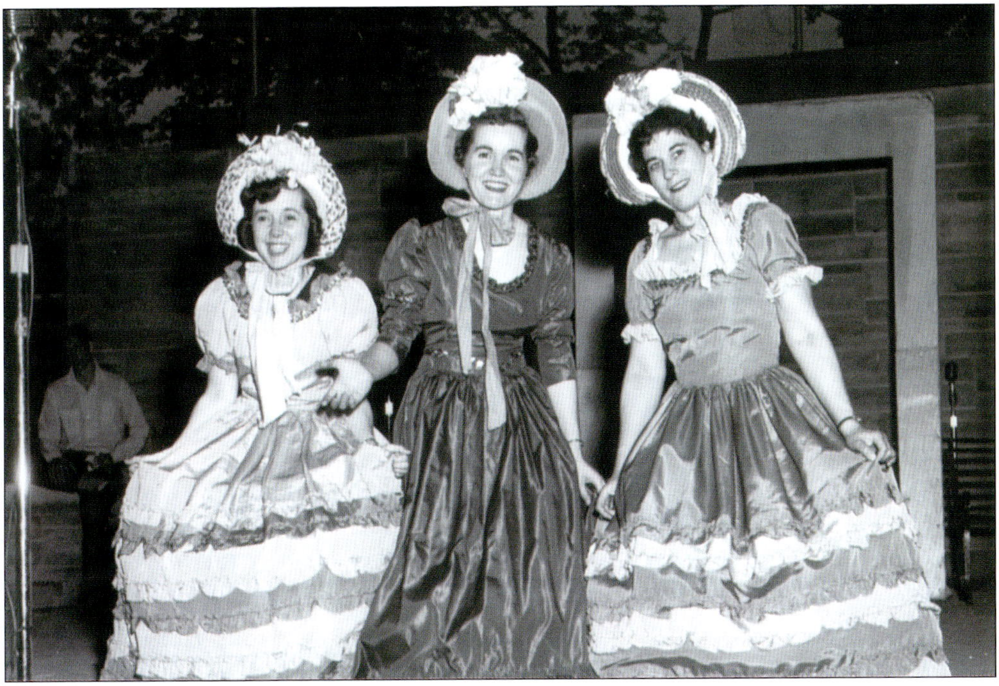

"Three Little Maids from School"—Gilbert and Sullivan's Mitzy, Fritzy, and Kitzy (above)—present quite a contrast to two acoustic guitarists from the Crawpuppies (below), both pictured in performance, though 70 years apart.

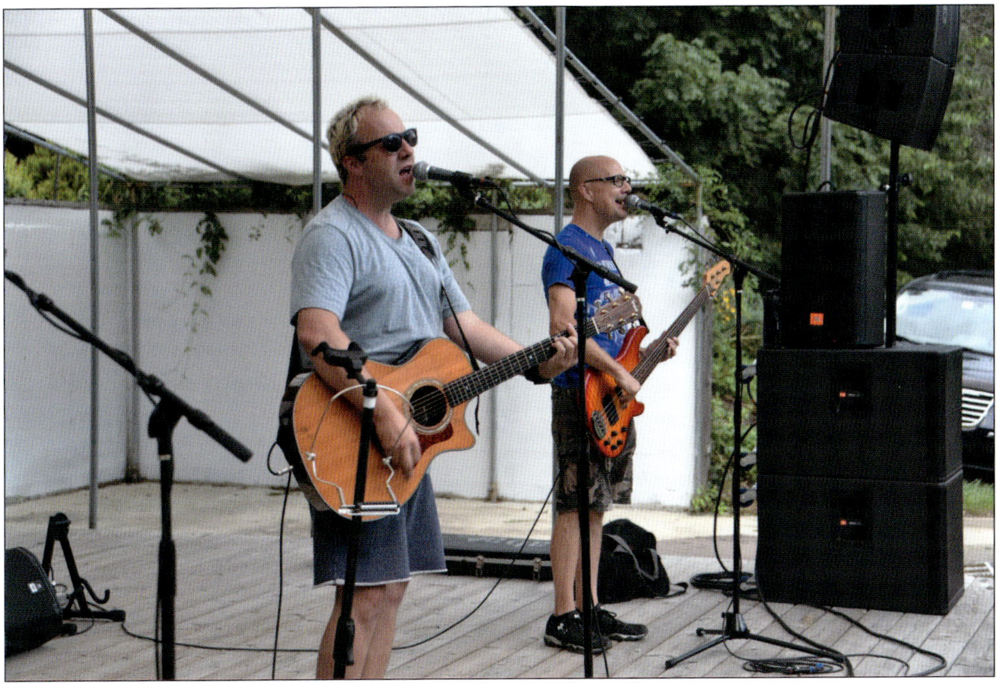

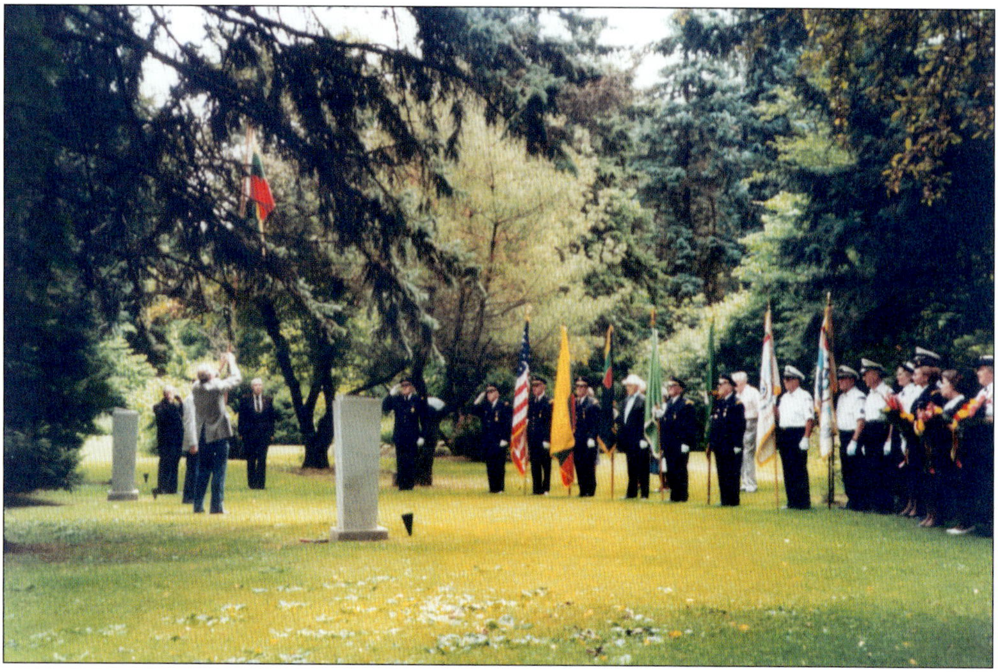

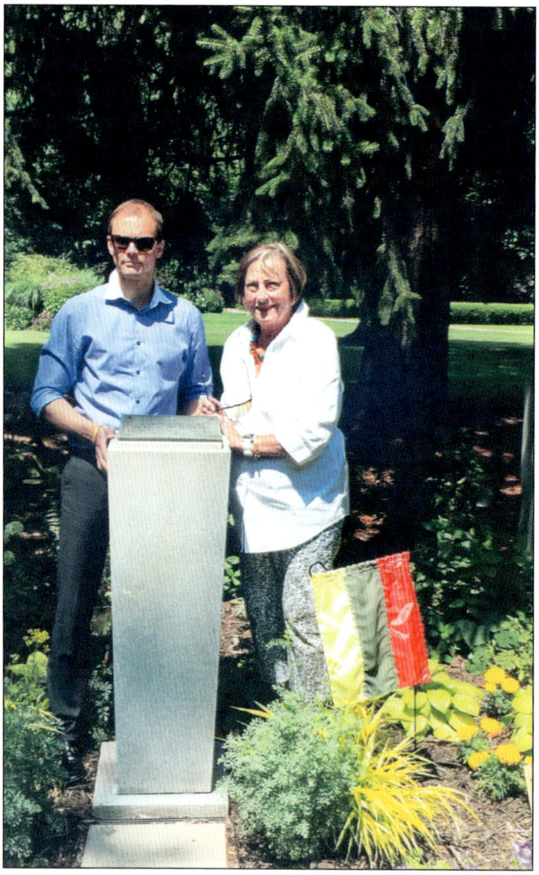

The Lithuanian gardens (above) hold a special significance for the Chicago-area Lithuanian community. In 1941, the first president of Lithuania, Antanas Smetona, came to International Friendship Gardens and planted a spruce tree to mark his country's independence following World War I. Karininku Ramove, the Lithuanian Military Officers Association, subsequently planted two additional trees, honoring Aleksandras Stulginskis, the second president, and Dr. Kazys Grinius, the third president. Commemorative ceremonies are pictured here. At left, in July 2017, Dr. Rima Binder (right) greeted Mantvydas Bekesius, Lithuanian consul general for Chicago and the Midwest. (Above, courtesy of Richard Houck; left, courtesy of Rima Binder.)

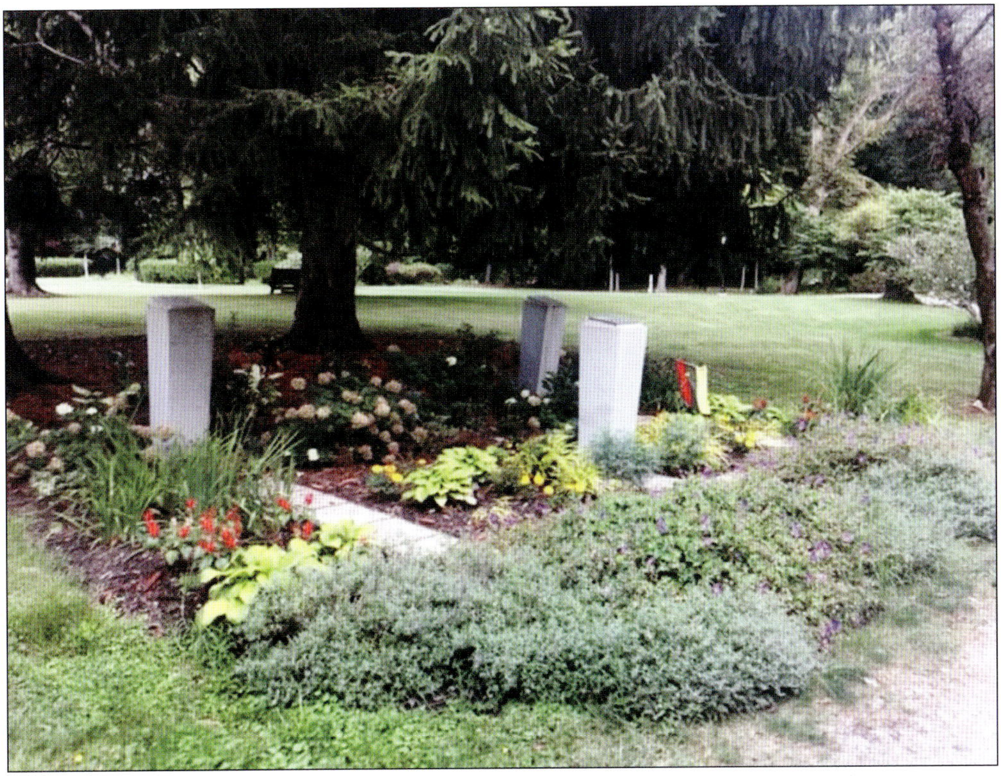

Above, stone monuments honoring Lithuanian presidents are installed beneath the spruce trees as memorials of this pre-Soviet period in the country's history. A Lithuanian wayside shrine stands in another part of the garden (below). Both areas are maintained by Dr. Rima Binder, a second-generation volunteer for Lithuanian causes. Binder, a resident of Long Beach, Indiana, is a master gardener, educator of gifted children, and president of the Lithuanian University Women's Association (LUMA). (Both, courtesy of Rima Binder.)

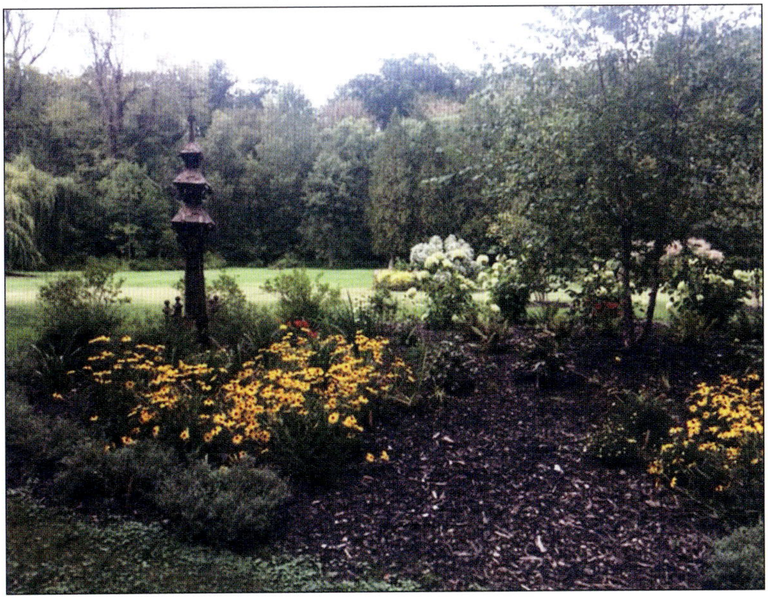

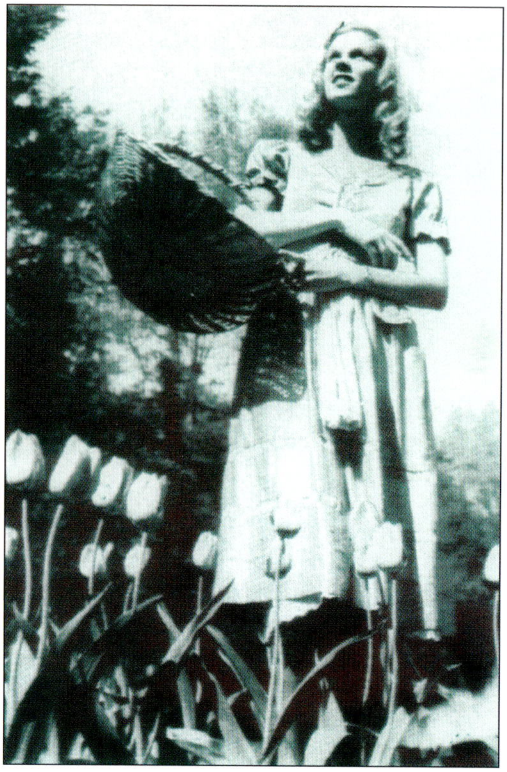

Way back then, a winsome blonde lady was posed among the tulips with a basket looped over her arm, suggesting she was soon to be gathering flowers. The blonde lady pictured below in 2017 is a working woman, Triecia Roberts, business manager of Friendship Botanic Gardens, who obligingly posed in front of this giant sycamore tree to illustrate the tree's great size. Estimated to date from 1818, the sycamore measures 13 feet in circumference. (Left, photograph by Tom Grieger; below, photograph by Phil Eichas.)

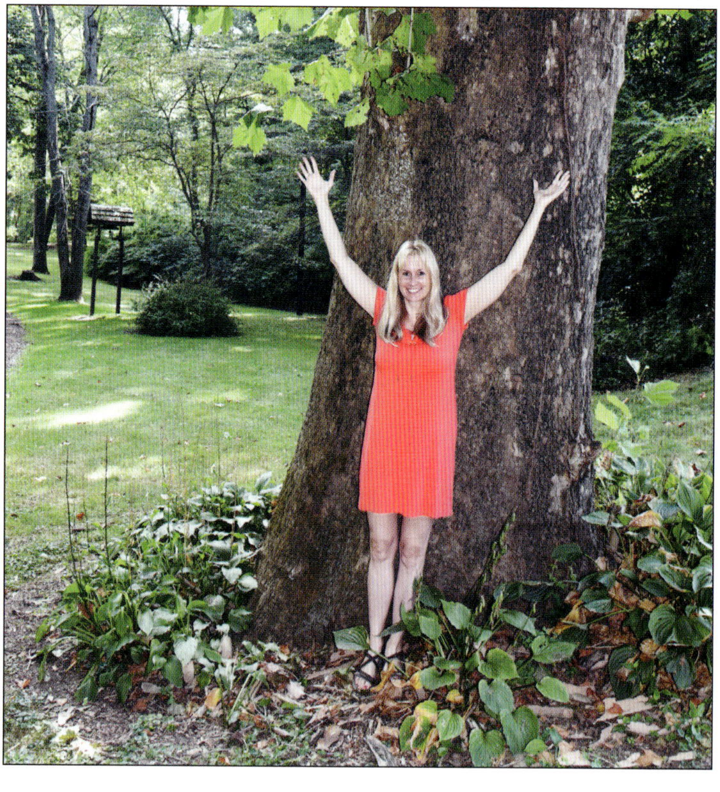

Operatic singers have appeared in the gardens for many years, spanning the decades from Lucille Manners (right) to a group from the Lyric Opera of Chicago (below). They are, from left to right, Rose Guccioni, soprano; Gary Alexander, narrator; Kenneth Donovan, tenor; Rima Binder, cochairman of the 2017 event; Kevin Keys, baritone; John Goodwin, pianist; and Janet Mensen Reynolds, mezzo-soprano. (Below, photograph by Phil Eichas.)

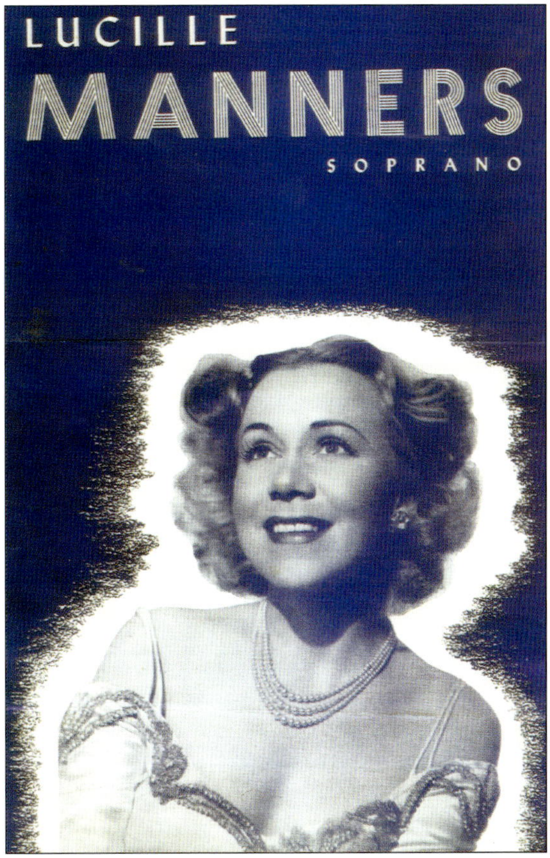

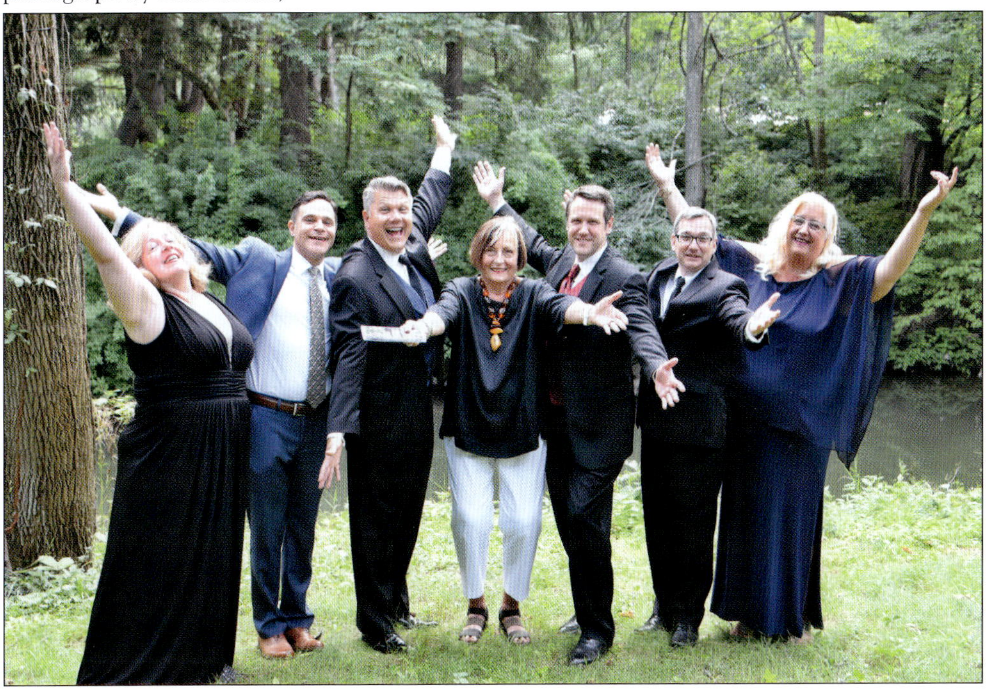

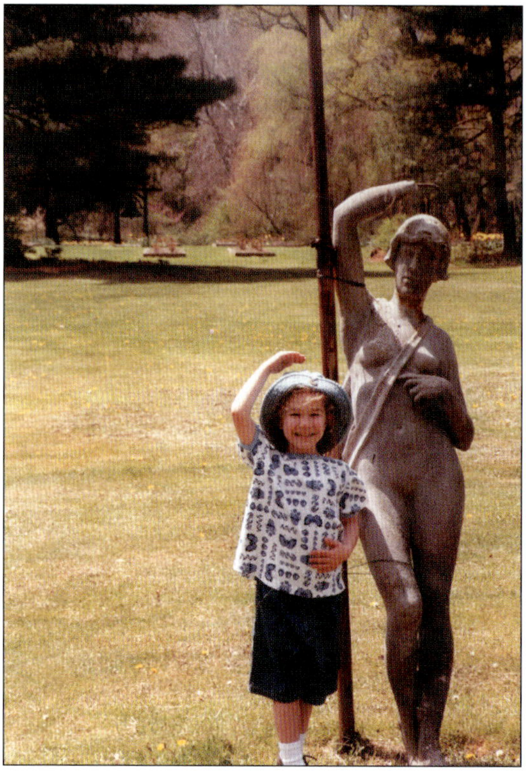

The naked lady statues have often inspired mischievous fun, as when the ballet dancer's pose is parodied by Tammy Steinhagen's young daughter Marque Steinhagen (left). Somebody decided the seated figure below needed a vest; whether for warmth or modesty was not reported. (Left, courtesy of Richard Houck; below, photograph by Phil Eichas.)

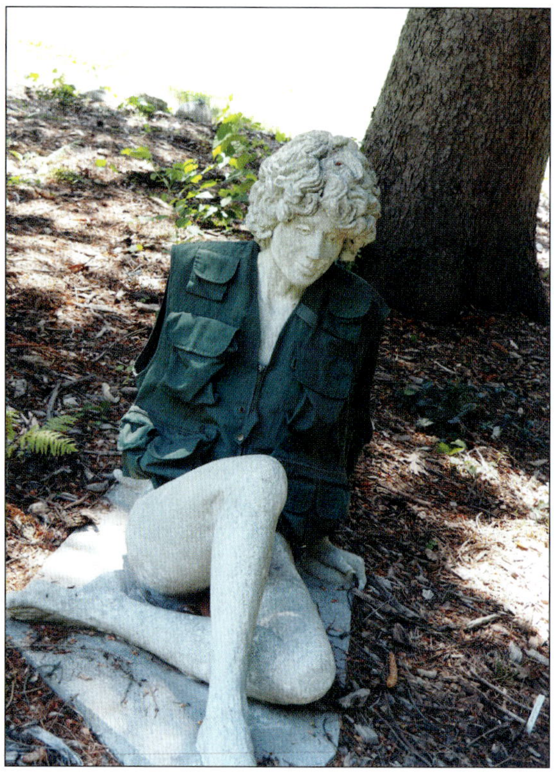

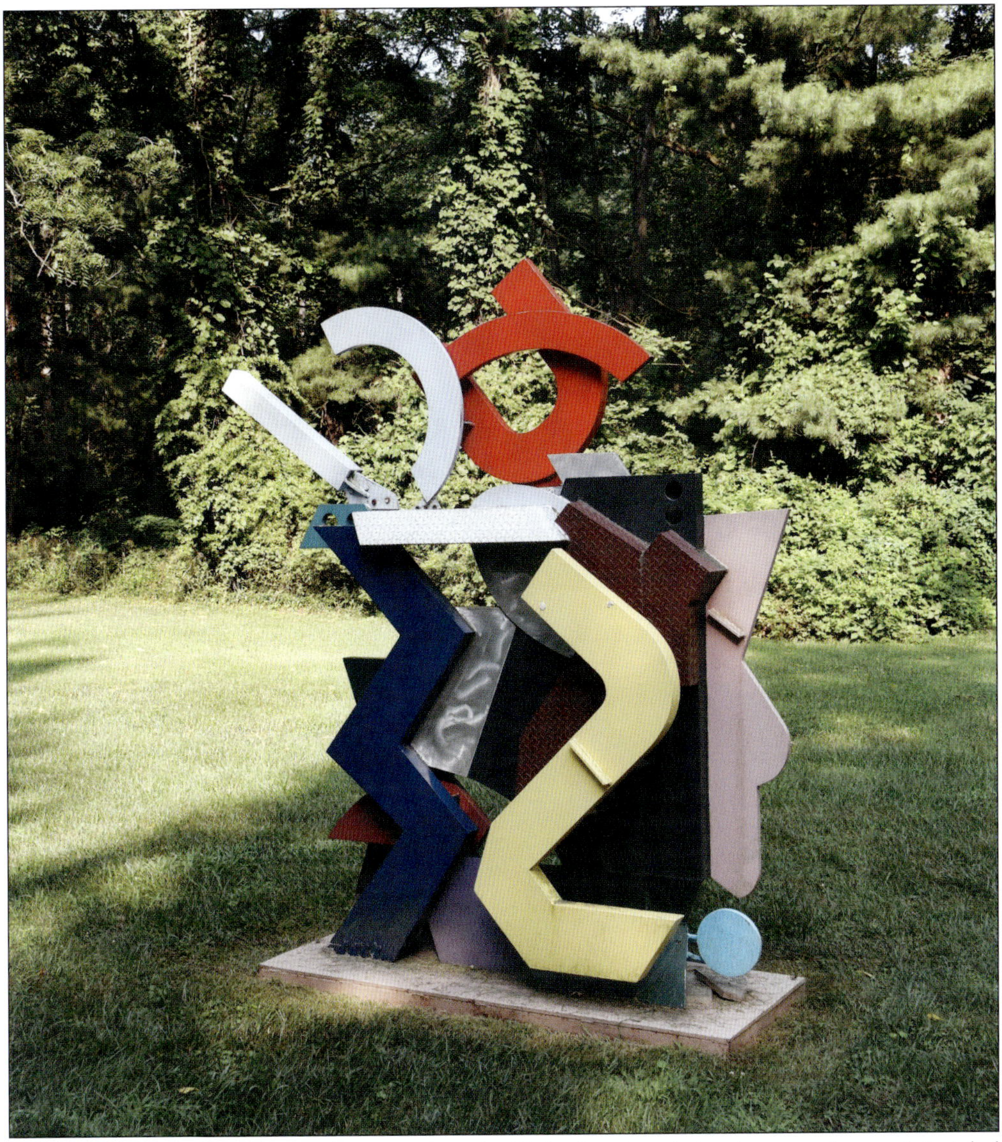

Sculptors today prefer abstract forms, such as Joe Dahm's painted aluminum sculpture. It is titled *Cubist Still-Life With Figure*, and a close look reveals subtle references to his female model. Dahm, a former resident of Chicago, Illinois, and Pottawattamie Park, Indiana, now lives in Port Townsend, Washington. Another one of his sculptures stands outside a Michigan City park department building. This piece was donated by the John Leinweber and James Laughlin family and stands in the ArcelorMittal Children's Garden. (Photograph by Phil Eichas.)

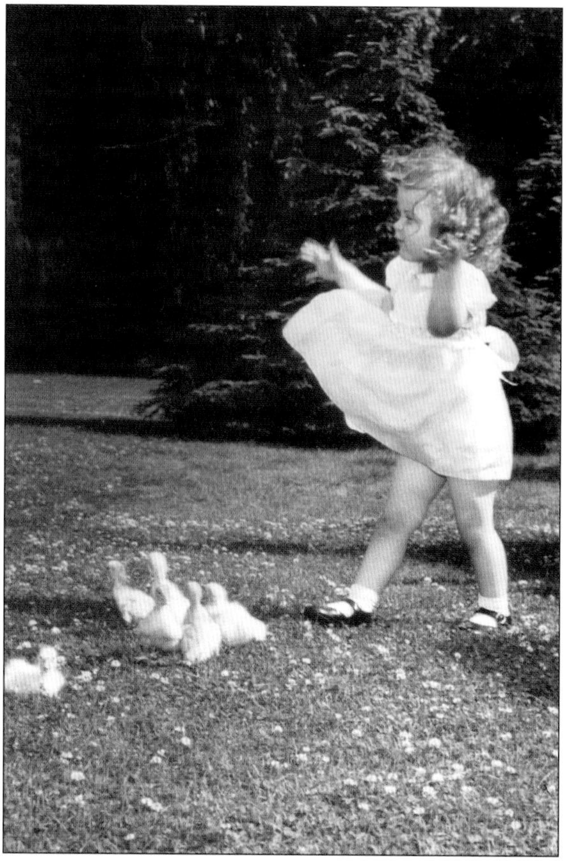

At left, a young girl chases baby chicks—an Easter Sunday photo op. Dipping little ducklings in egg dye was one amusement of the 1970s, not the sort of thing fostered in today's eco-friendly world. Below, baby turtles find a safe and sunny habitat at Friendship Botanic Gardens. (Below, photograph by Phil Eichas.)

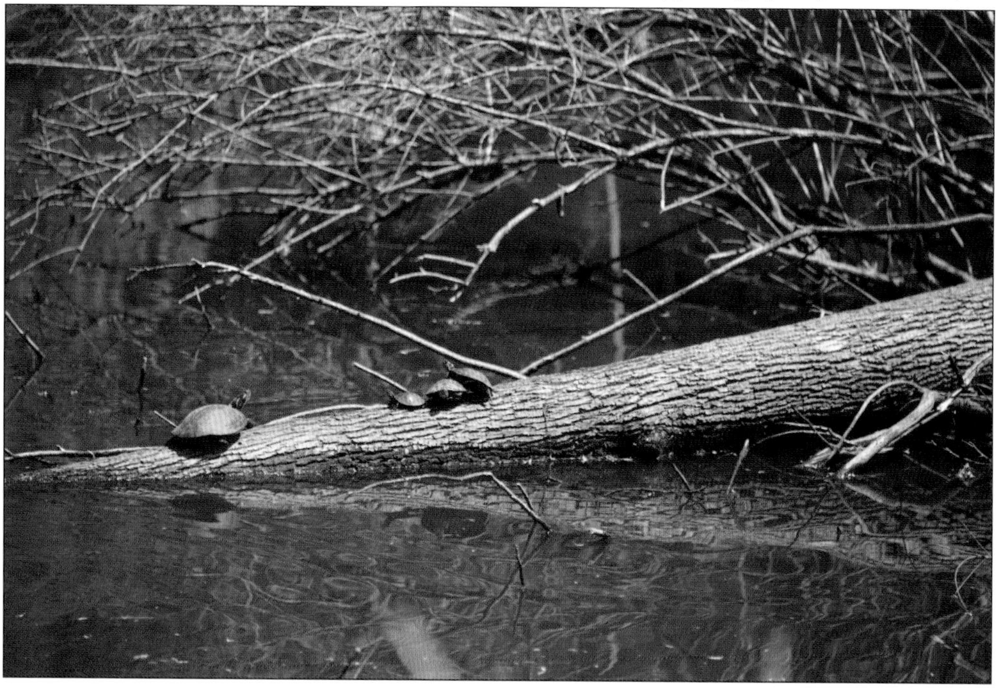

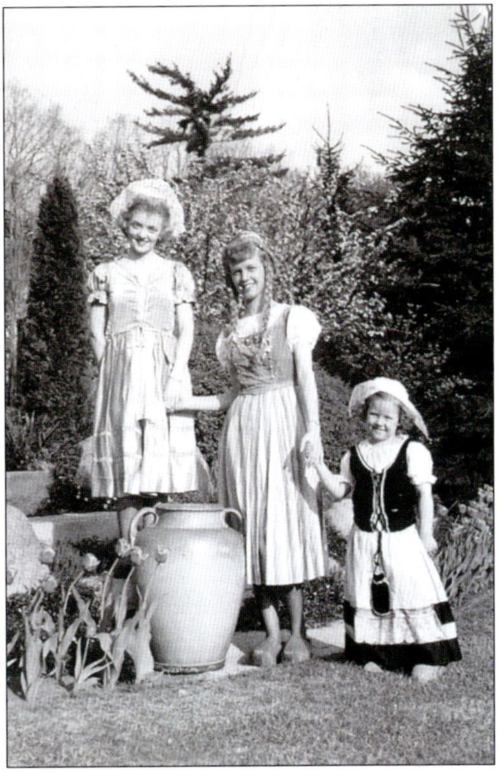

International Friendship Gardens brought out the old-country fashions of several European groups, none photographed more often than Dutch girls with their braided blonde hair and wooden shoes (right). Young girls still love dressing up for costume parties, and Halloween at Friendship Botanic Gardens gave that opportunity to, from left to right (below), Ellie Shuck, Anna Kate Mereness, and Sabrina Isaacson (Below, photograph by Phil Eichas.)

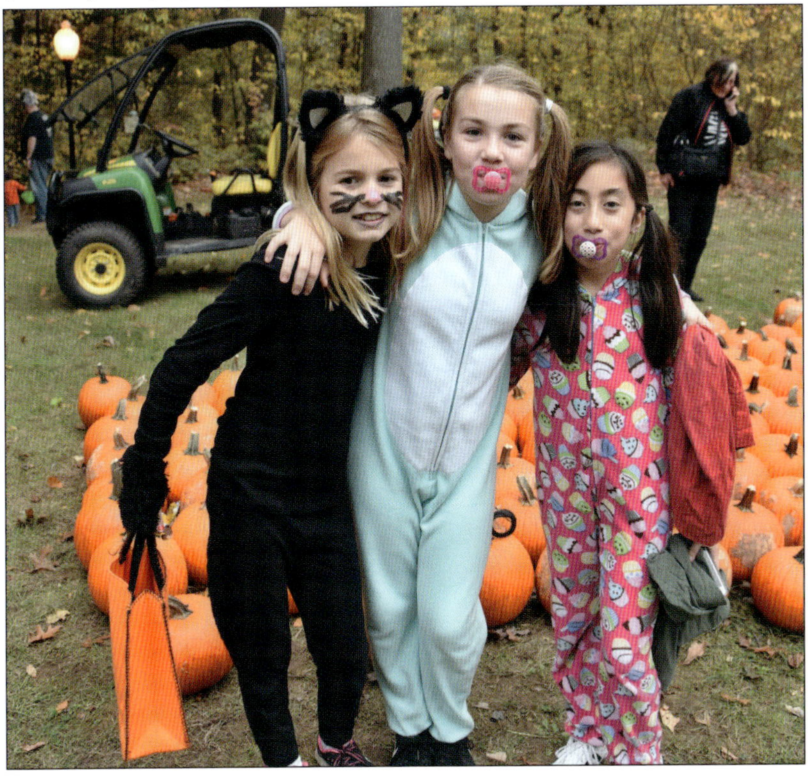

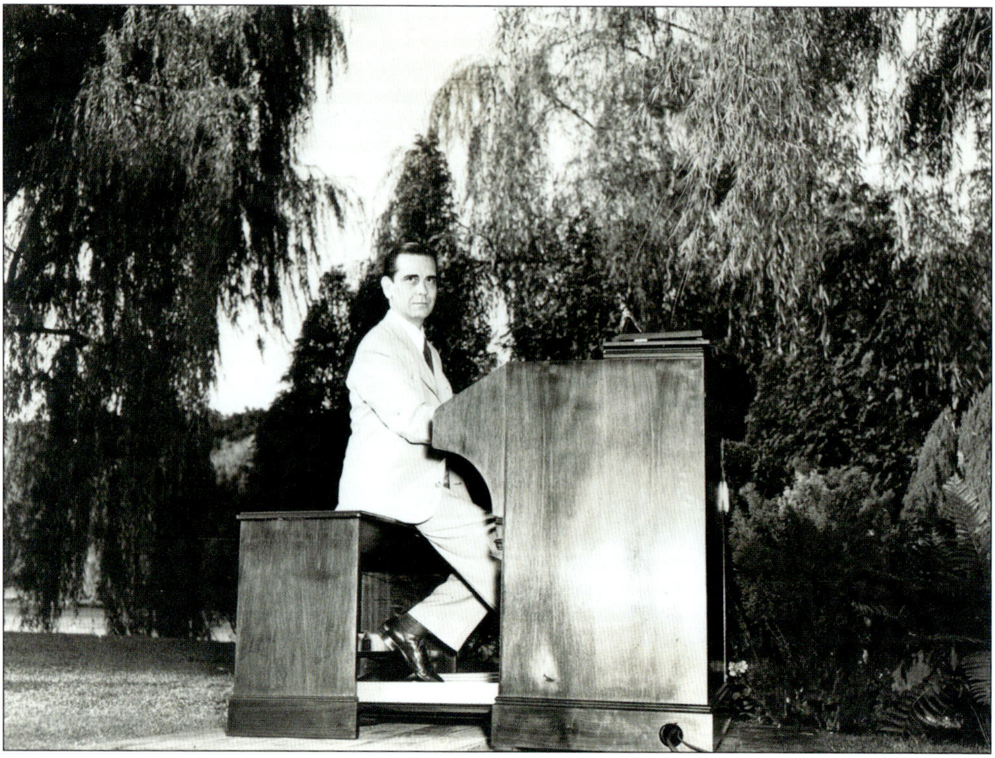

The willow tree setting pictured is unmistakably International Friendship Gardens, and the penciled inscription on the back of the c. 1945 photograph above identifies the musician as Dr. Bethuel Gross, rehearsing at a Hammond organ—and taking this event very seriously. A 2016 performance, by contrast, shows the Crawpuppies drummer letting loose on his drums. (Below, photograph by Phil Eichas.)

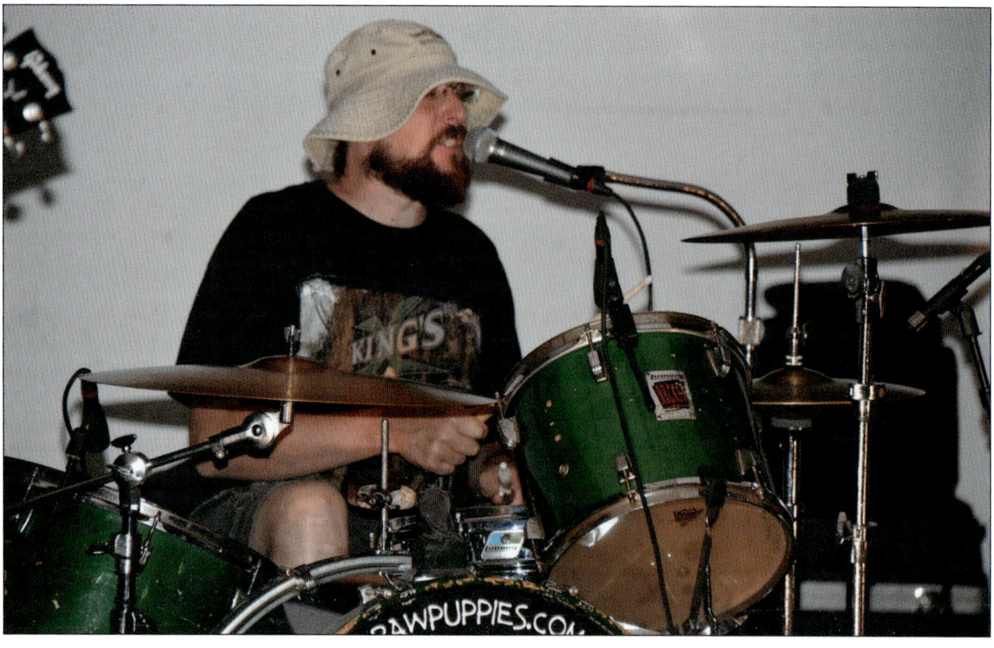

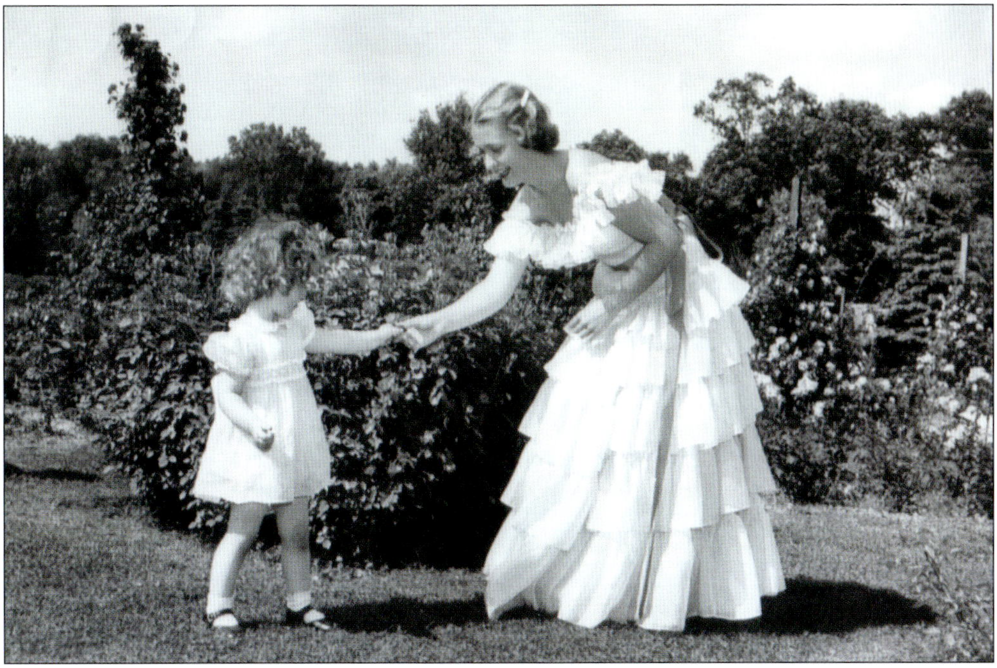

What a difference! A gracious lady (above) in her ruffled gown around 1948 is coaxing the curly-haired youngster, dressed like a Shirley Temple doll, into her garden of earthly delights—an obvious allusion to a growing-up experience. In today's world, young girls gain center stage, sitting patiently while a volunteer face painter gives a quick makeover for an evening of Halloween make-believe. The message is similar but differences can be perceived, both in their attire and the interaction between generations. (Above, photograph by Tom Grieger, courtesy of Friendship Botanic Gardens; below, photograph by Phil Eichas.)

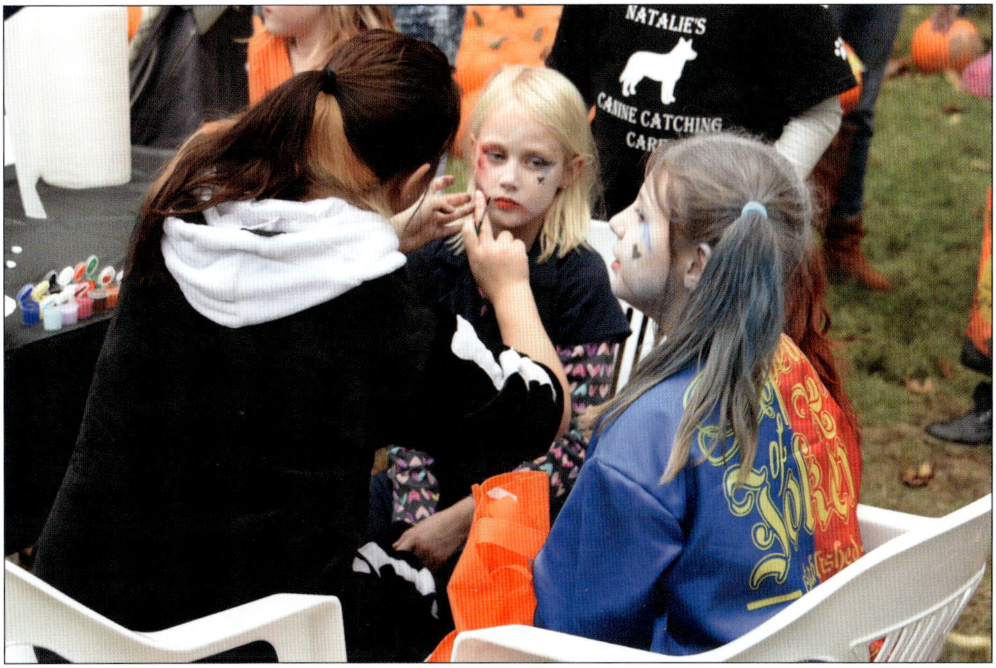

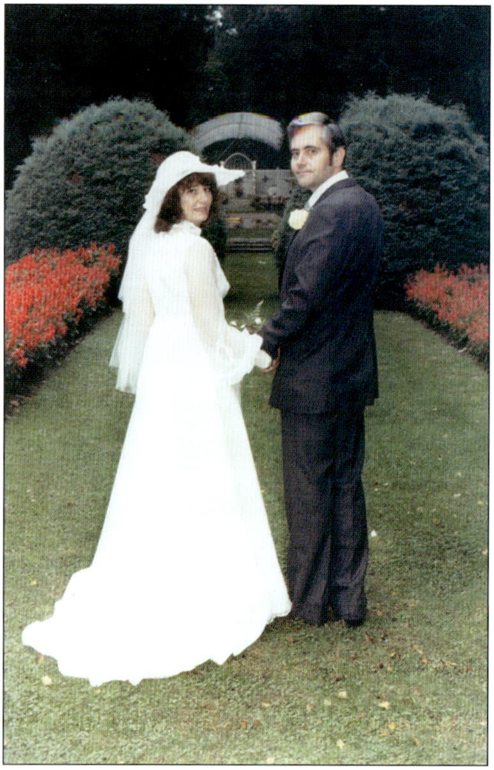

Mary Fox and Bruce Hoppe held their wedding ceremonies at International Friendship Gardens in 1981 and were photographed in Symphony Garden. (Courtesy of Mary Fox.)

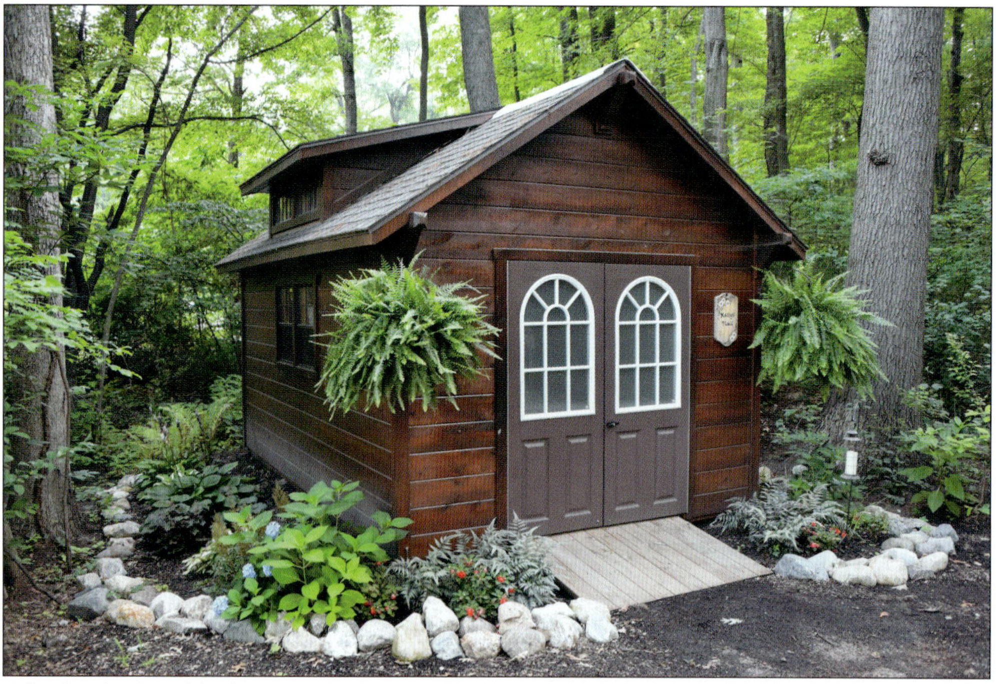

After her husband died 30 years later, Mary Fox had a small cabin built in his memory. It is called Mary's Place and serves as a dressing room for the bride and her attendants. Fox is an attorney and current board member at Friendship Botanic Gardens. (Photograph by Phil Eichas.)

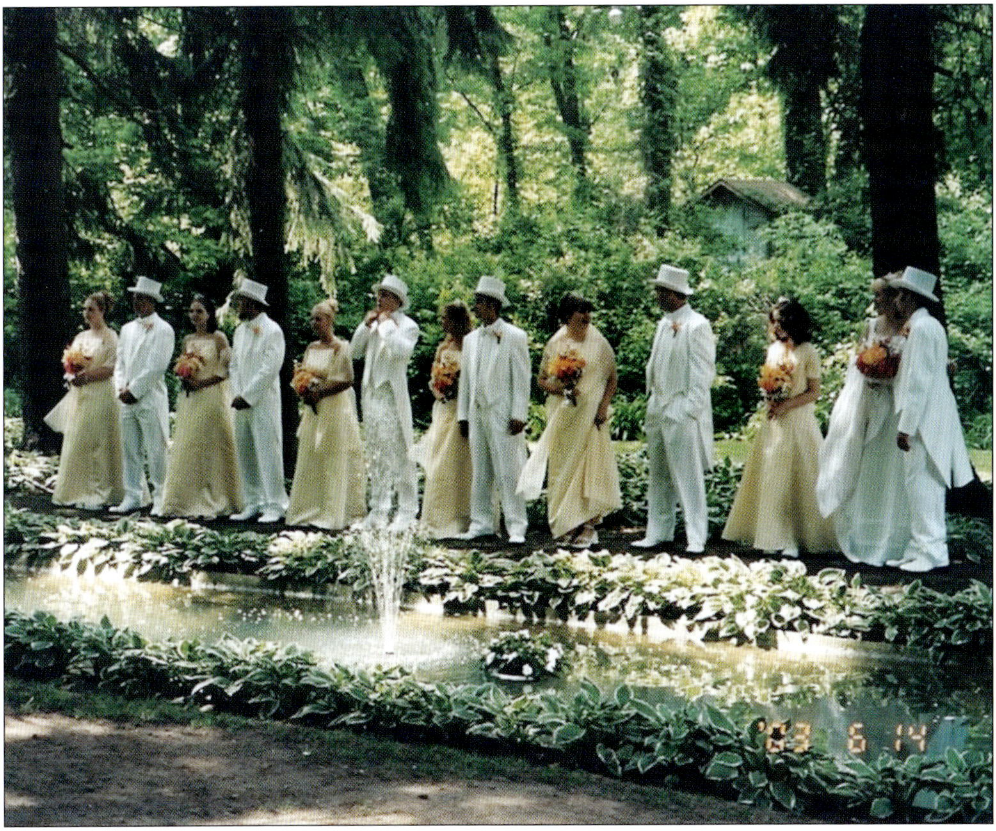

Weddings have taken place in these gardens for more than 50 years. Once again, they provide a stunning background for the bride and groom and their stylishly dressed attendants. (Both, courtesy of Friendship Botanic Gardens.)

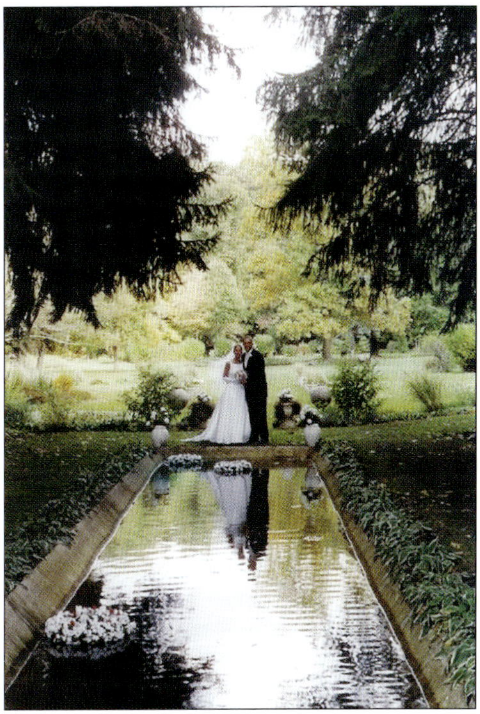

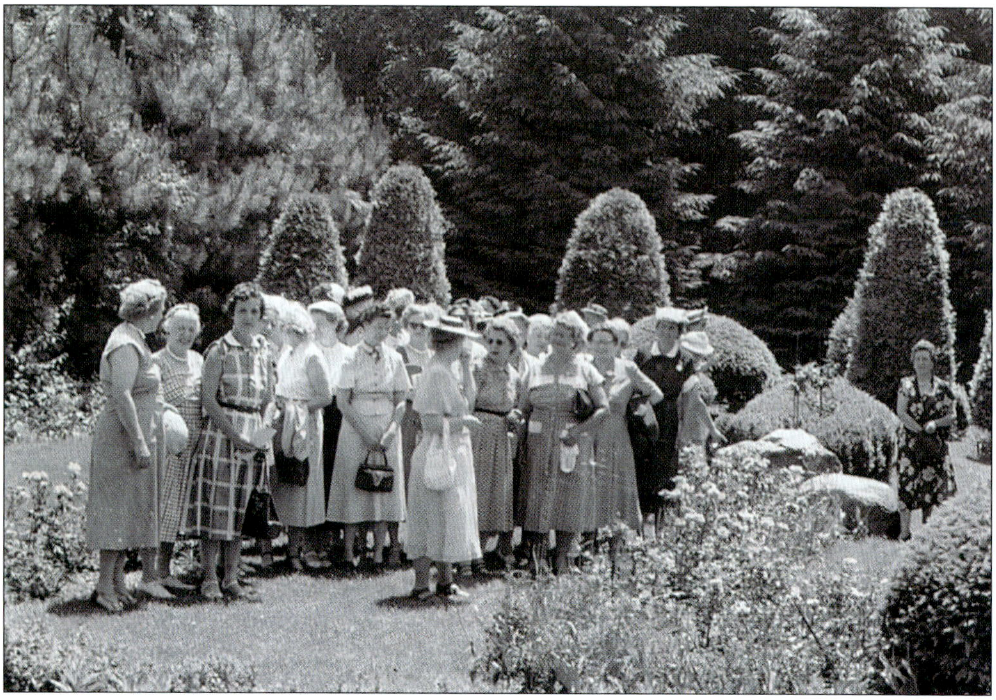

The differences between then and now. Above, a group of demurely dressed ladies in the late 1950s have apparently departed from their tour bus and are awaiting a guided tour of the gardens. Below, family groups await instructions for capturing butterflies, an event initiated in 2017 and actively involving moms, dads, and lots of eager kids. (Below, photograph by Phil Eichas.)

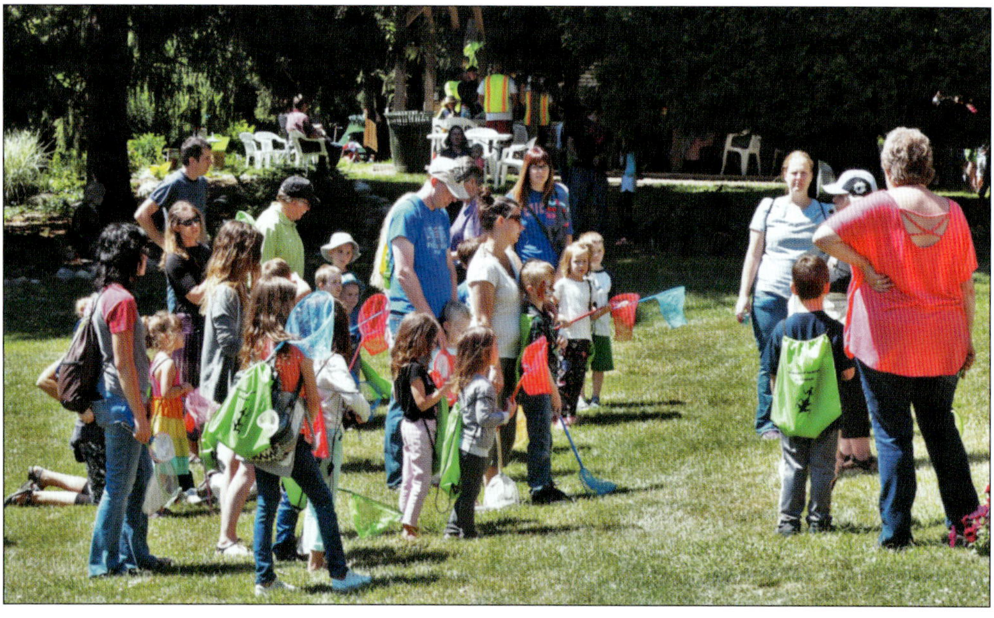

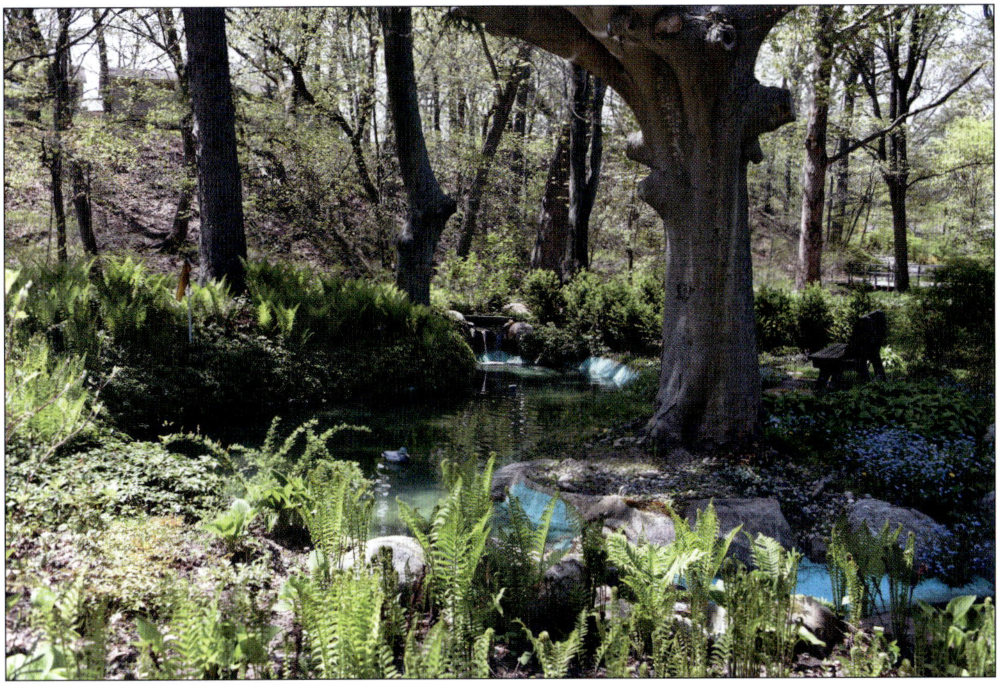

As it was written in the original contract between Dr. Frank Warren and the Stauffer brothers and as it remains today, half of the land was to be left in its original state (above) and the other half to be cultivated as gardens (below). Adding to the original 100 acres, the Stauffers later purchased five more acres, which included frontage on US Highway 12. (Both photographs by Phil Eichas.)

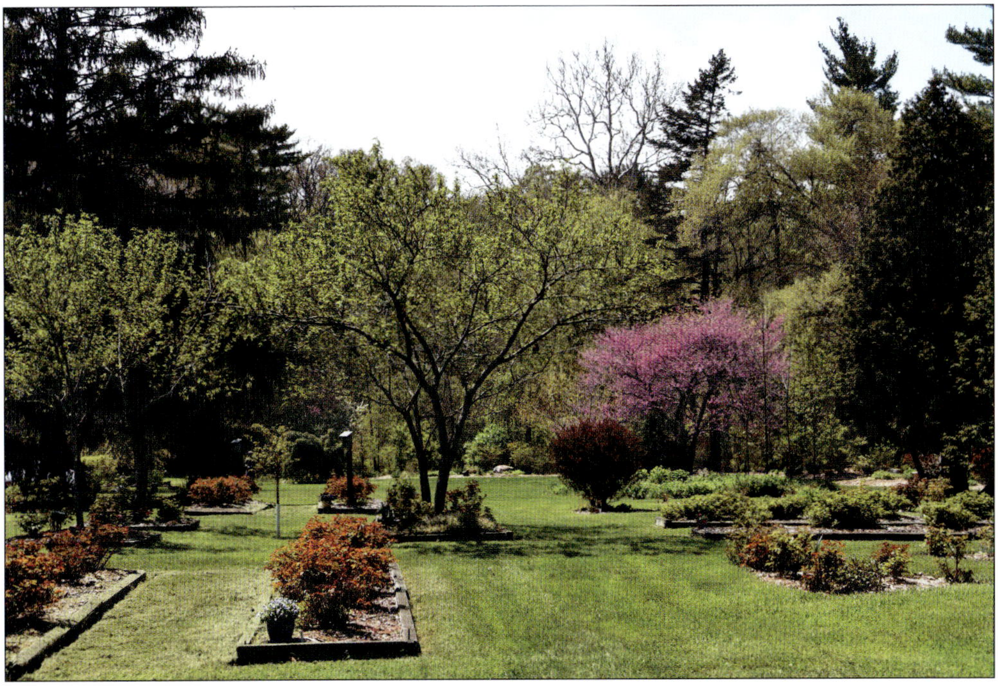

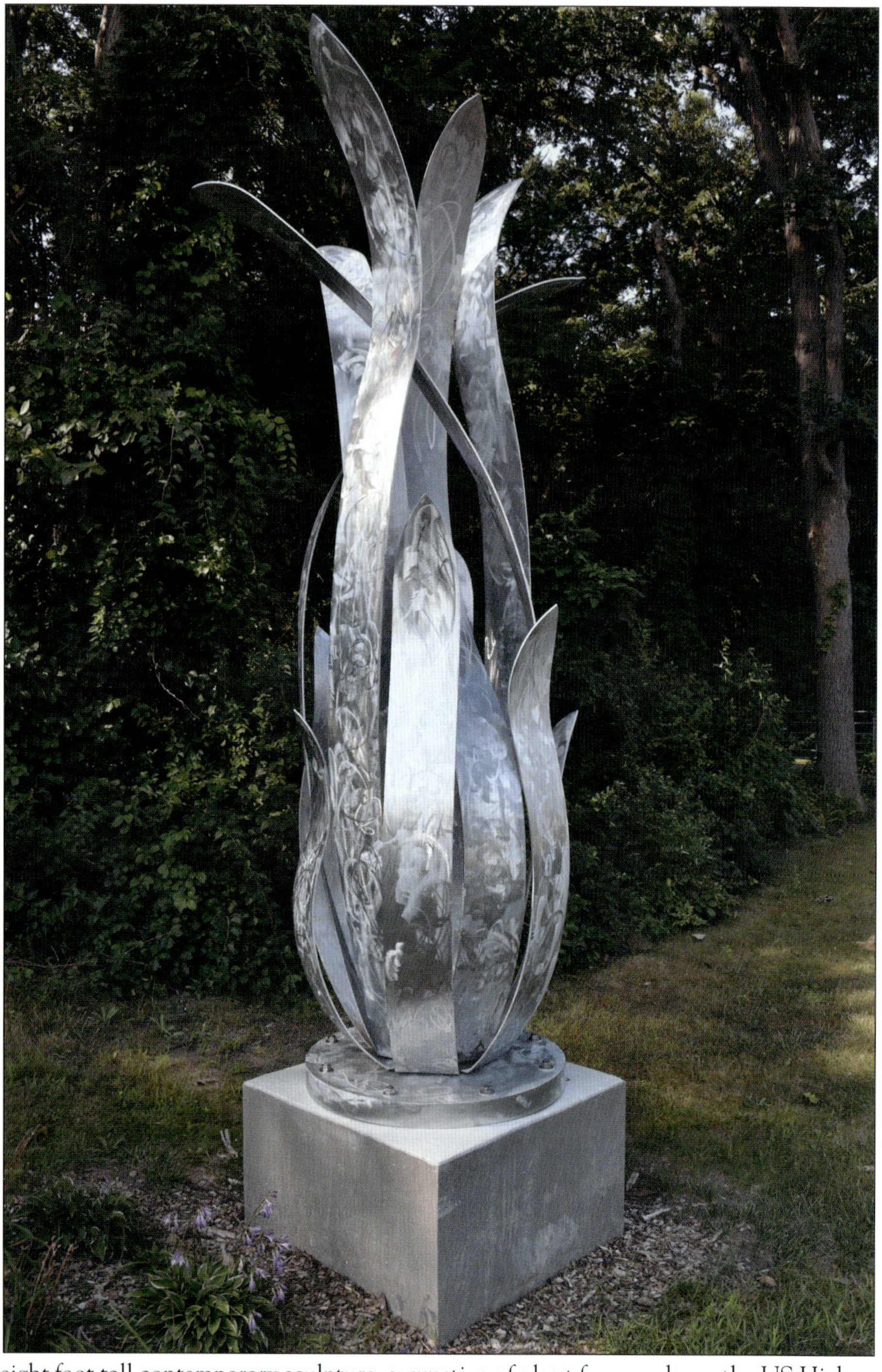

An eight-foot-tall contemporary sculpture, suggestive of plant forms, adorns the US Highway 12 entrance to Friendship Botanic Gardens in Michigan City, Indiana. Entitled *Botanic Magic*, the eye-catching piece is made of silvery aluminum. John Searles, the sculptor, lives about 10 miles away, in Harbert, Michigan. The sculpture was donated to the gardens by the John Leinweber and James Laughlin family.

Six

A New Beginning

Nobody can say exactly when the new beginning began, but everybody can agree it happened before it was too late. At least one family fortune was spent trying to create and maintain International Friendship Gardens. Then, when the tended gardens were overrun by invasive plants and costs exceeded income by large amounts, funding became available from new sources.

In 1999, the year that Tammy Steinhagen became executive director, major grants came in from the Unity Foundation of La Porte County and Michigan City Community Enrichment Corporation. Steinhagen managed the gardens for six years. In 2005, Marti Pizzini was named executive director, and she was succeeded by Nancy Jahnel-Barnes, who remained in charge until 2015.

By that time, the rescue efforts had gained momentum. In 2001, International Friendship Gardens received a $156,794 grant from the Build Indiana Fund—"the result of unrelenting efforts," according to Dr. Houck, by Indiana state senator Anita Bowser. A longtime resident of Trail Creek, Bowser served in the legislature from 1992 until her death in 2007. The Build Indiana money was used to help fund a new, more visible entrance to the gardens on US Highway 12.

In 2014, John Leinweber became president of the board of directors. In addition to making substantial personal contributions, he developed new funding sources: ArcelorMittal, Horizon Bank, Barker Welfare Foundation, and Michigan City Redevelopment Commission. Other local companies and individuals saw the growth and development of the gardens and began to lend their support.

The greater emphasis on scientific and academic pursuits is reflected in the new name, Friendship Botanic Gardens. Innovative events, such as Bug Safari and Butterfly Bonanza, introduce children to nature and its importance. New types of group activities are refreshing the regular programming. Rotary Club revitalized its garden. Board member Kevin Egan and his son installed a Juvenile Diabetes garden. The Audubon Society conducts bird-watching tours. Greater use is made of hiking trails and fishing opportunities. The results are seen in attendance figures. In the past three years, attendance has increased from 4,000 to over 25,000. The gardens are beautiful and peaceful, a welcome escape from the hustle of traffic and technology.

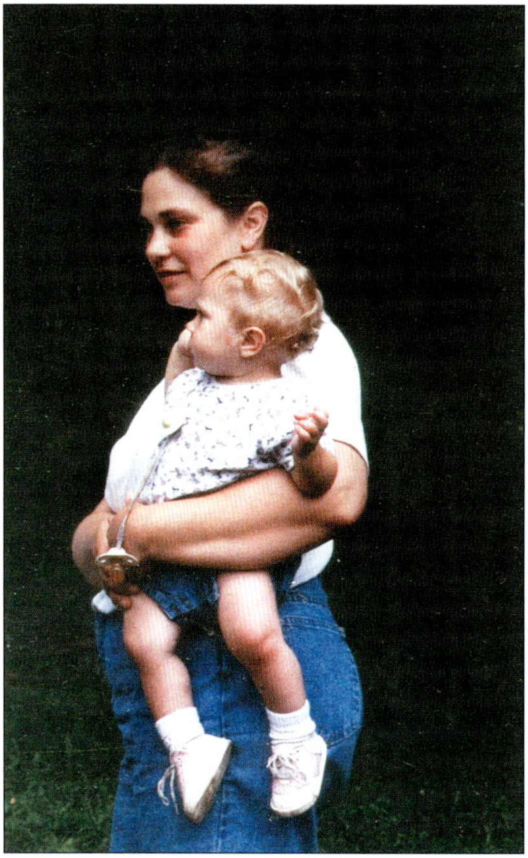

Tammy Steinhagen, described by colleagues as "the soul of the gardens," is pictured with her daughter at a special event in the 1990s. Steinhagen served as executive director from 1999 to 2005, after several years of volunteer work. She died at the age of 45 in 2012. (Courtesy of Richard Houck.)

On a gorgeous autumn afternoon, crowds of visitors line up at the entrance gate, patiently waiting for admission to the Halloween party, a highly popular event. Friendship Botanic Gardens provides a perfect setting for haunted trails, spooks in the woods, pumpkin painting, and other amusing activities.

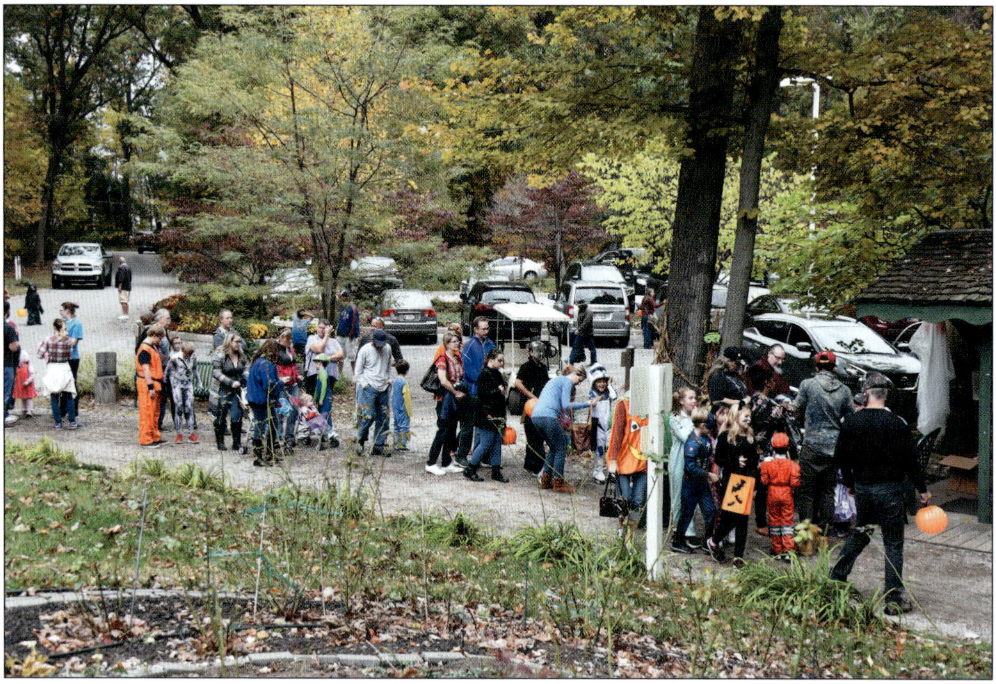

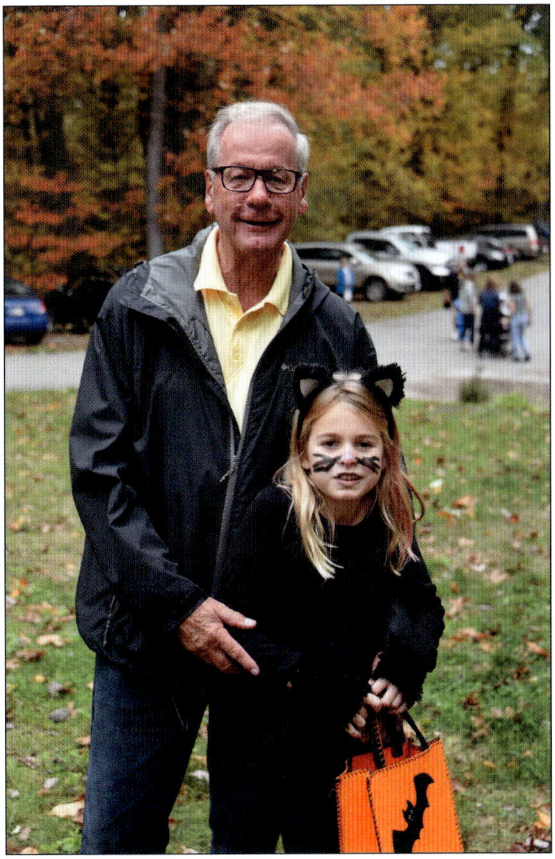

John Leinweber (right), board president of Friendship Botanic Gardens, attends Halloween festivities in 2016 with his granddaughter Ellie Shuck. Rima Binder (below), board vice-president, accompanies her two grandsons Titas Binder (left) and Lukas Binder (right). The Halloween party is one of the intergenerational festivities that the current board of directors supports to increase family attendance. (Both, courtesy of Triecia Roberts.)

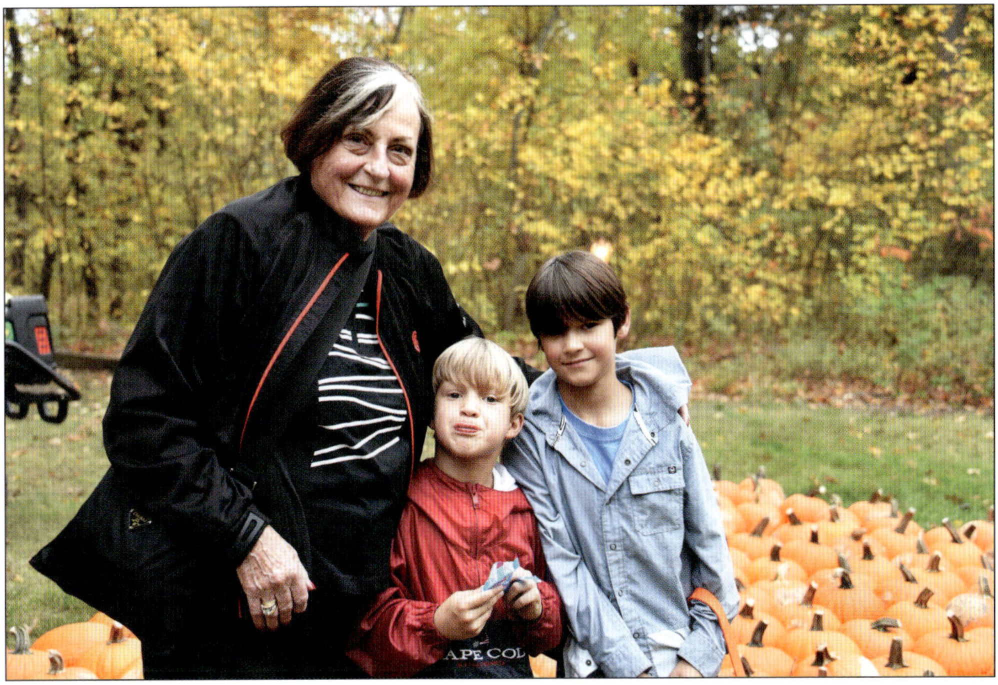

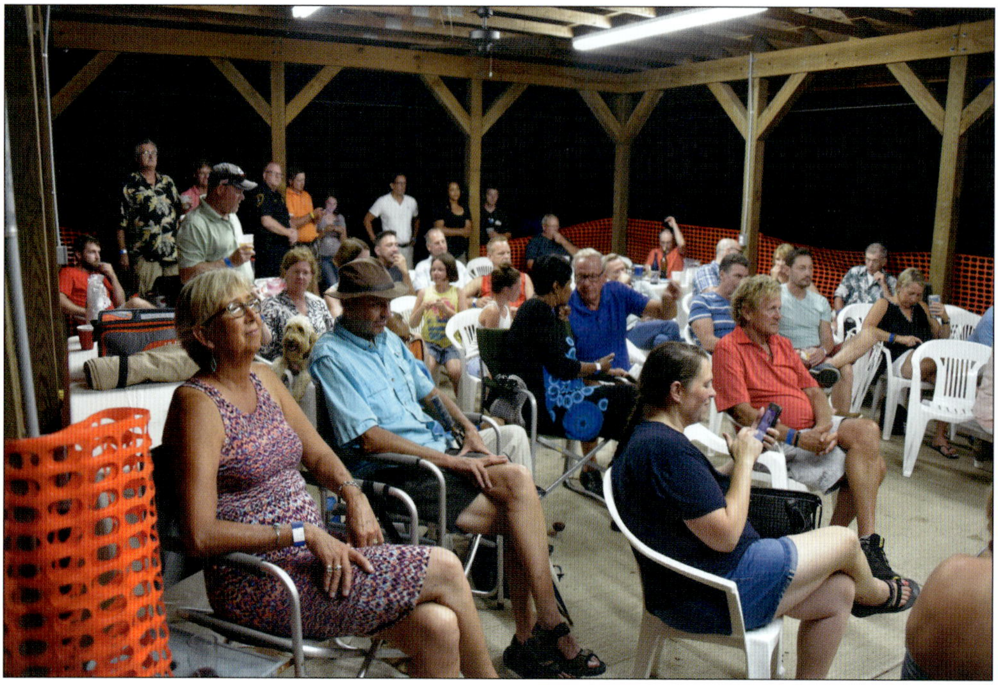

An attentive audience (above) listens to the music of Karisa Wilson, an award-winning singer, songwriter, and guitarist from Grand Rapids, Michigan (pictured below). Named "Best New Artist of the Year" in 2008, Wilson performed indie folk and jazz and blues at Friendship Botanic Gardens in the summer of 2016.

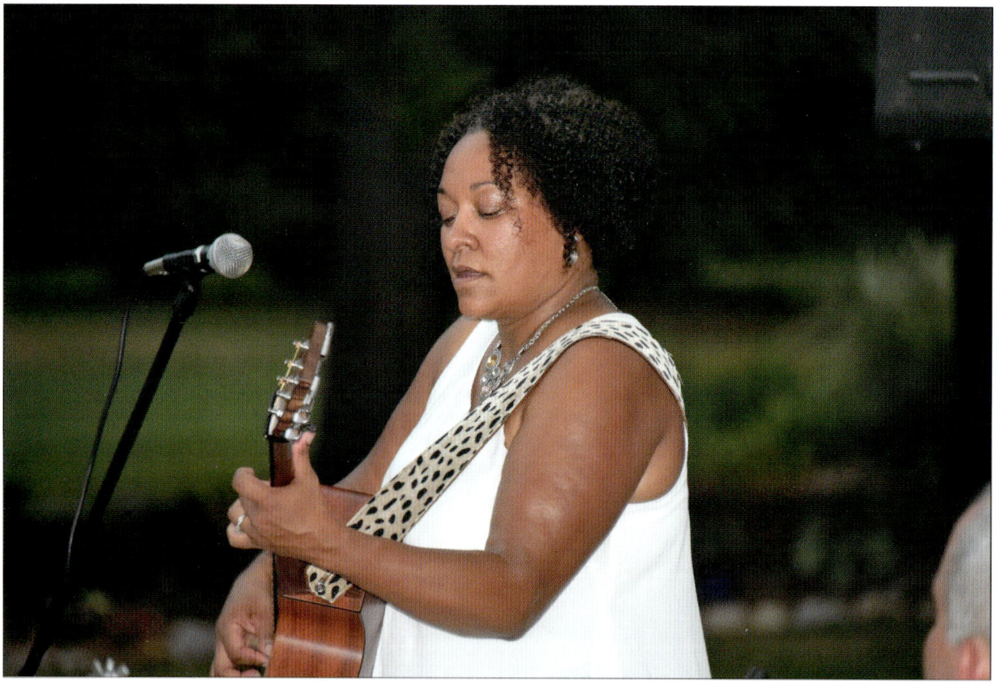

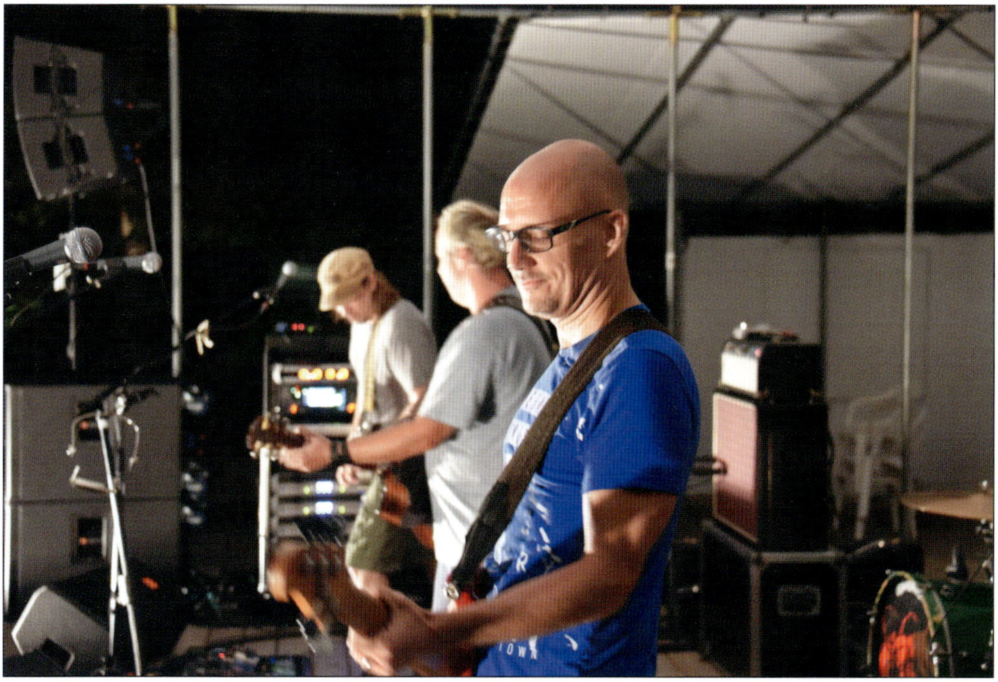

Aaron Hedges (right) plays guitar with the Crawpuppies, a pop rock band from Valparaiso, Indiana, shown in performance at the gardens in 2017.

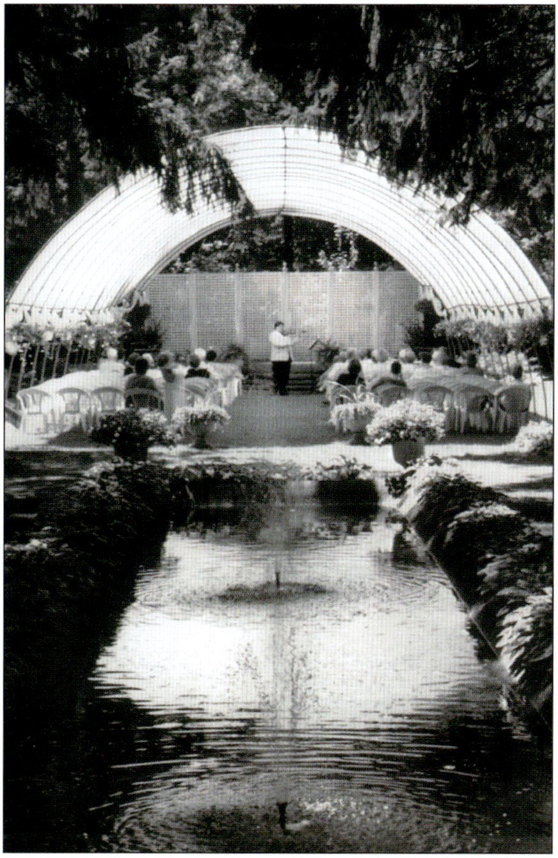

An evening of classical music in Symphony Garden featured Nic Orbovich, a Michigan City, Indiana, violinist who plays with the South Bend Symphony Orchestra. Orbovich is the founder of the highly acclaimed Michigan City Chamber Music Festival, now in its 15th year.

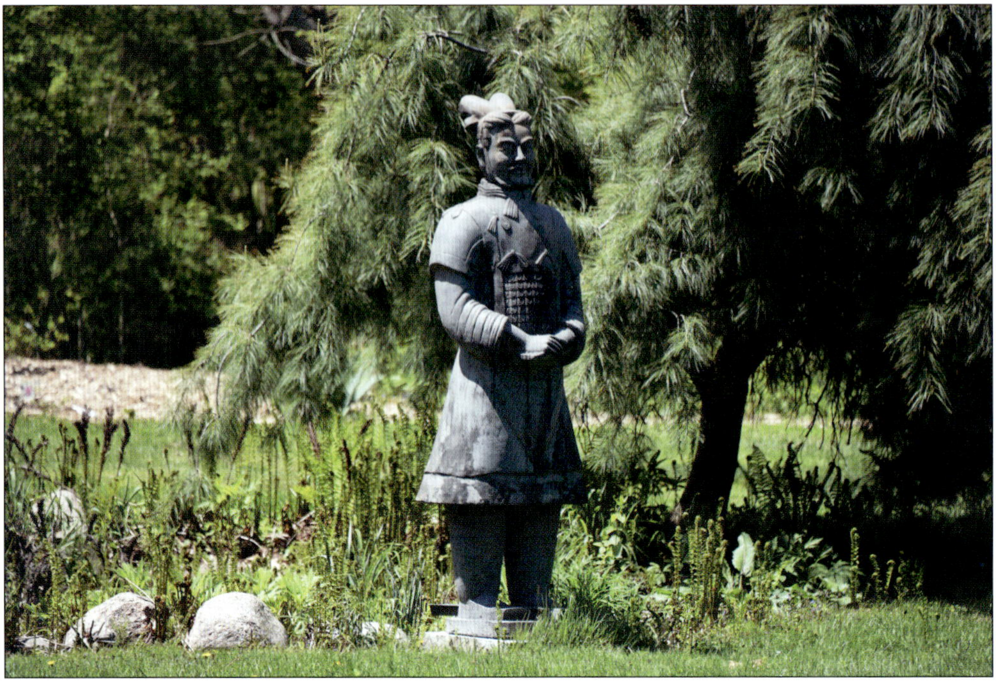

The life-sized Chinese warrior sculpture (above) illustrates the blend of tradition and happy coincidence that enlivens the history of the gardens. The terra-cotta figure was a gift to Ann Dahm, then president of the Michigan City Chamber of Commerce, from the owner of a company that conducts tours of China; Dahm then presented it to Friendship Botanic Gardens. The warrior is a replica of a Qin dynasty figure created centuries ago to guard the tomb of the first emperor of China. It was unearthed in 1974, in Shaanxi province, by farmers digging a well. The half-moon bridge (below), a longtime survivor of the original gardens, is now enhanced by a field of rudbeckia (black-eyed Susans).

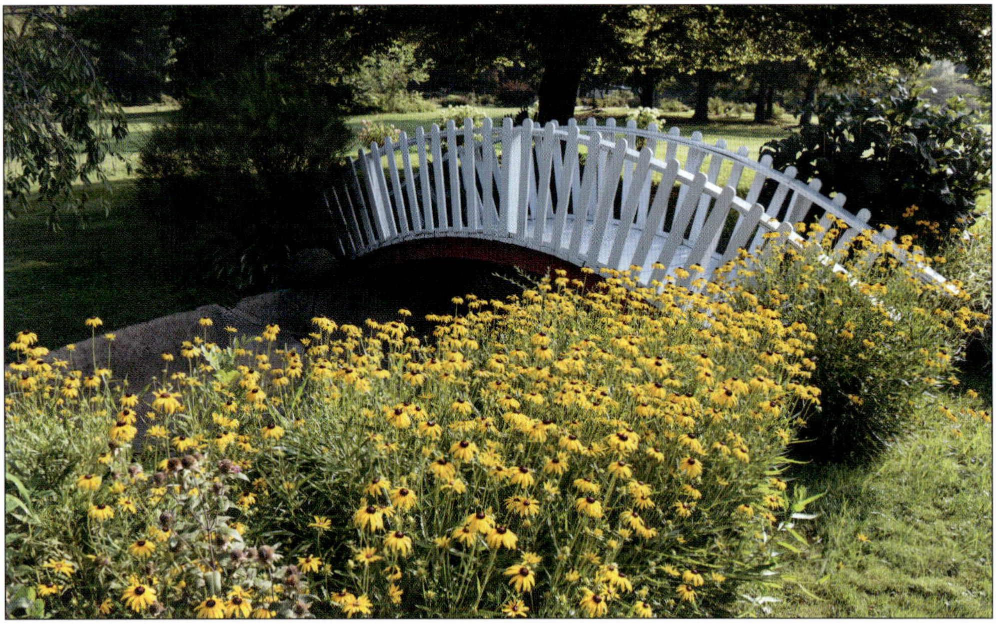

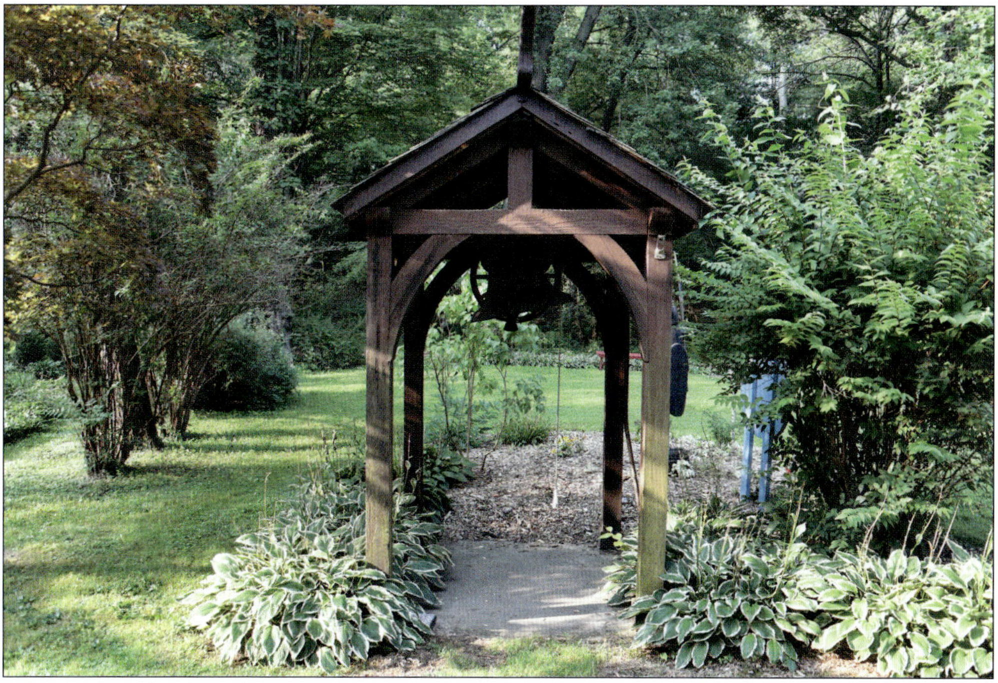

A prominent feature of the Norwegian garden (above) is an old church bell from Dowagiac, Michigan, acquired when Joseph Stauffer was searching for large bells to enhance International Friendship Gardens. Olief Olsaker (right) is pictured with his wood carving of a Viking adventurer. The Norwegian gardens are tended by Olsaker and his wife, Marianne, both native Scandinavians.

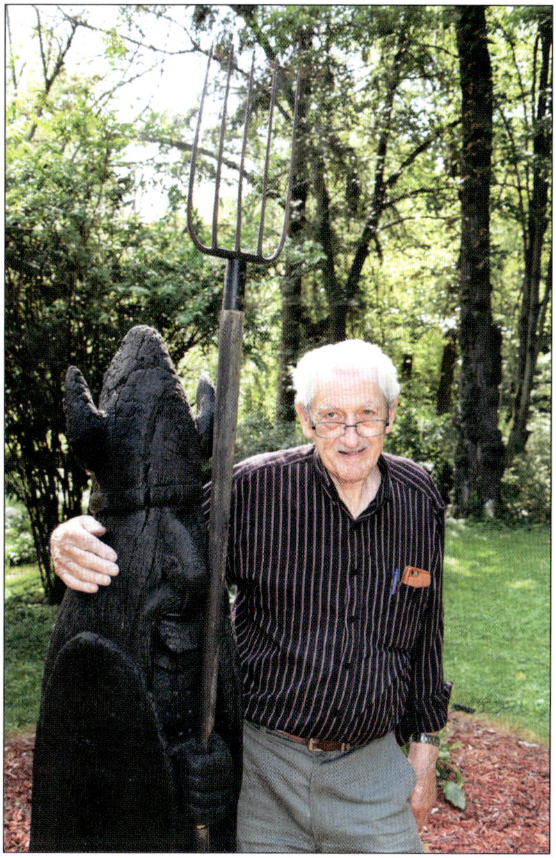

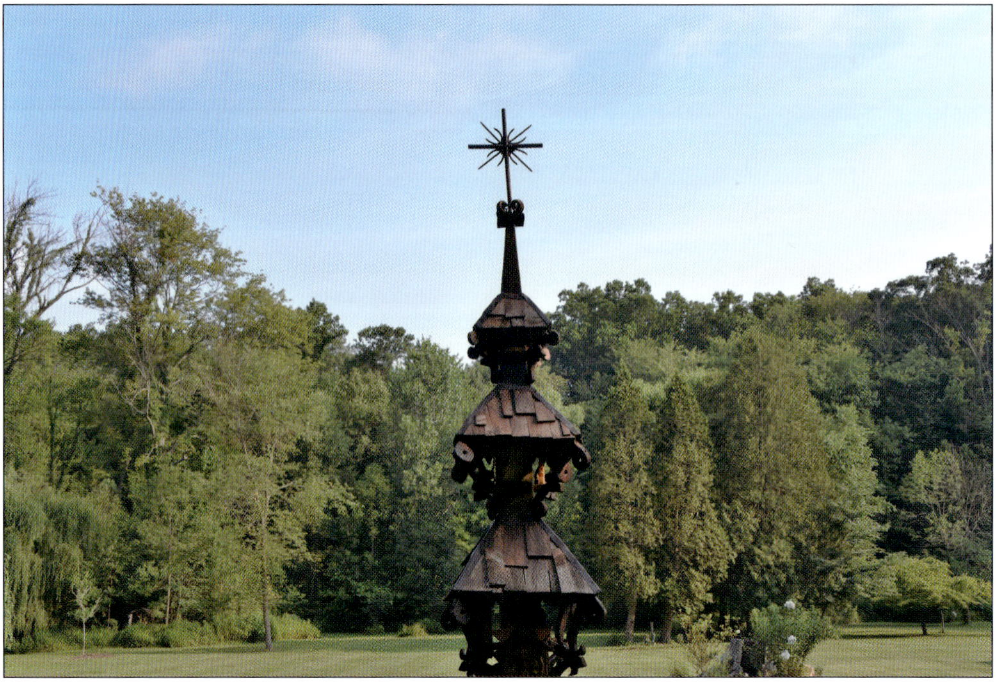

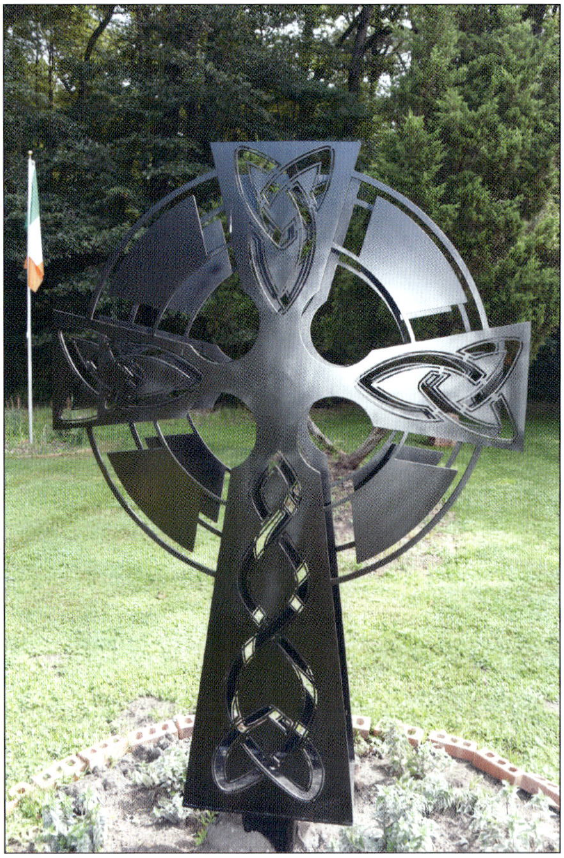

Rising above flower beds, the Lithuanian folk sculpture combines angels and other Christian emblems into a post or pole type of wood carving, topped by a cross. This is the typical design for wayside shrines seen in the villages of Lithuania. Dr. Rima Binder arranged for Friendship Botanic Gardens to acquire this piece created by Dr. Romuoldas Povilaitis of Lemont, Illinois, through a donation from the Knights of Lithuania Mid-America District.

A Celtic cross and flag of Ireland have been placed in the Irish garden being developed by Johnny Stimley, a Michigan City businessman.

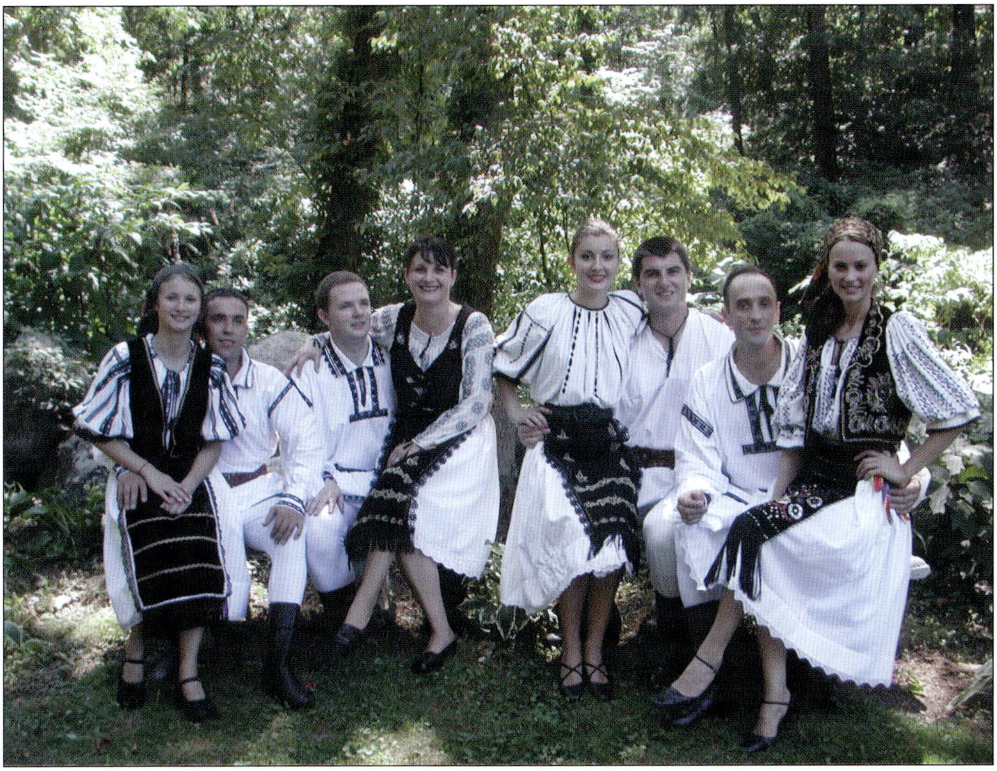

A group of Romanian folk dancers, shown in their embroidered costumes, are pictured after a recent performance at Friendship Botanic Gardens.

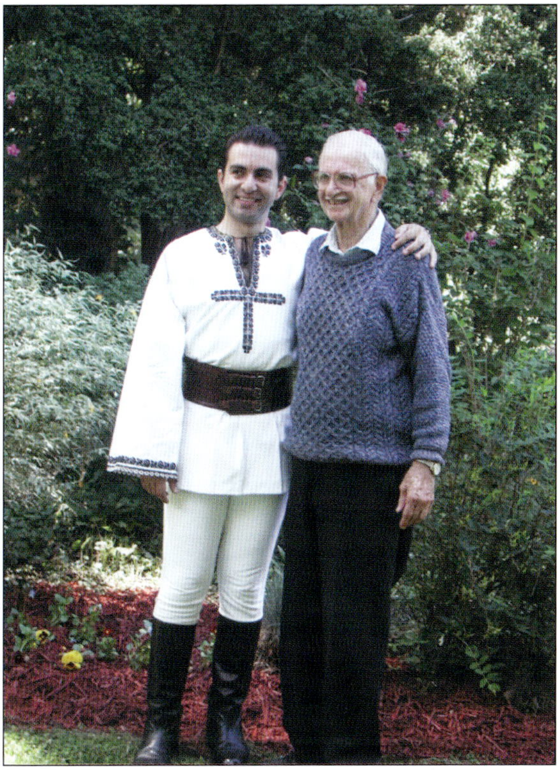

One of the dancers poses with George Neagu, who created the Romanian gardens and assisted with the Polish gardens and festivals prior to his death in 2017. The son of Romanian immigrants, Neagu was a child welfare worker, educational psychologist, founder of the Michigan City Human Relations Commission, and a member of many other social justice organizations.

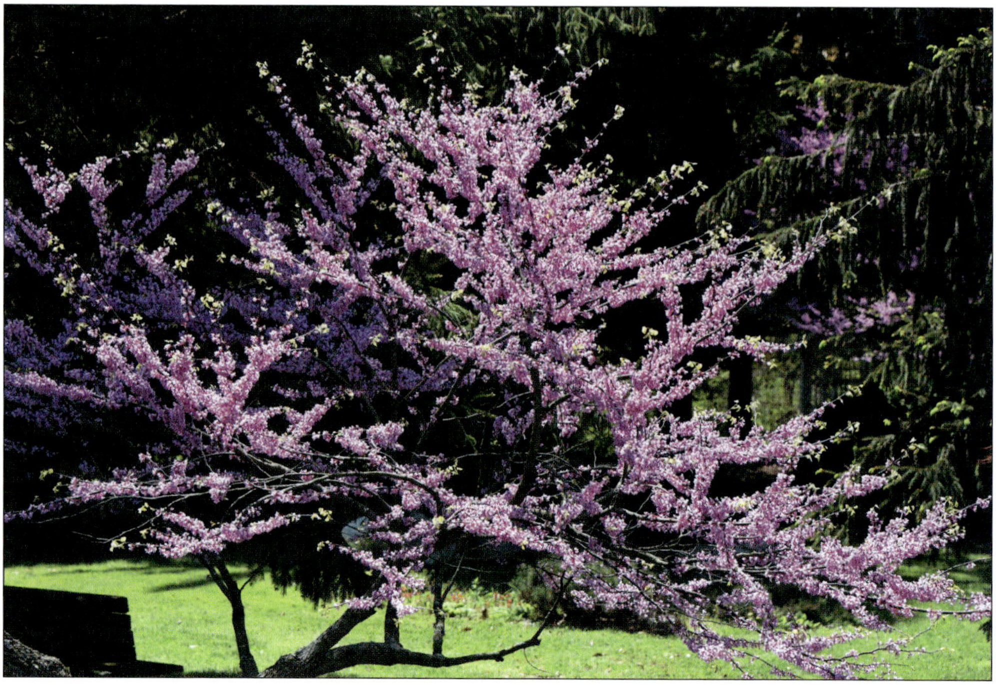

Colorful in all seasons, Friendship Botanic Gardens displays the glorious blooms of the redbud tree in springtime (above) and a swallowtail butterfly in mid-summer (below).

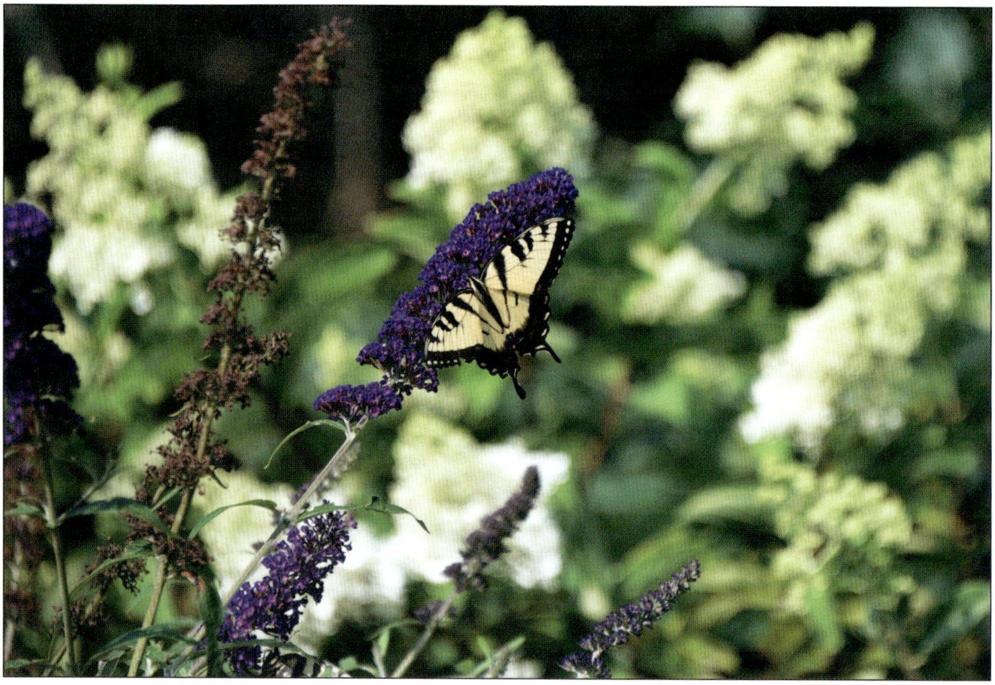

With greater usage of the trails, Friendship Botanic Gardens have begun extending their activity season. On Thanksgiving Day, a Turkey Walk and scavenger hunt attract the before-dinner crowd, as depicted humorously by Michigan City artist Neil Kienitz (right). Snowfall has tempted some members to use the trails for cross-country skiing. (Below, courtesy of Marianne Olsaker.)

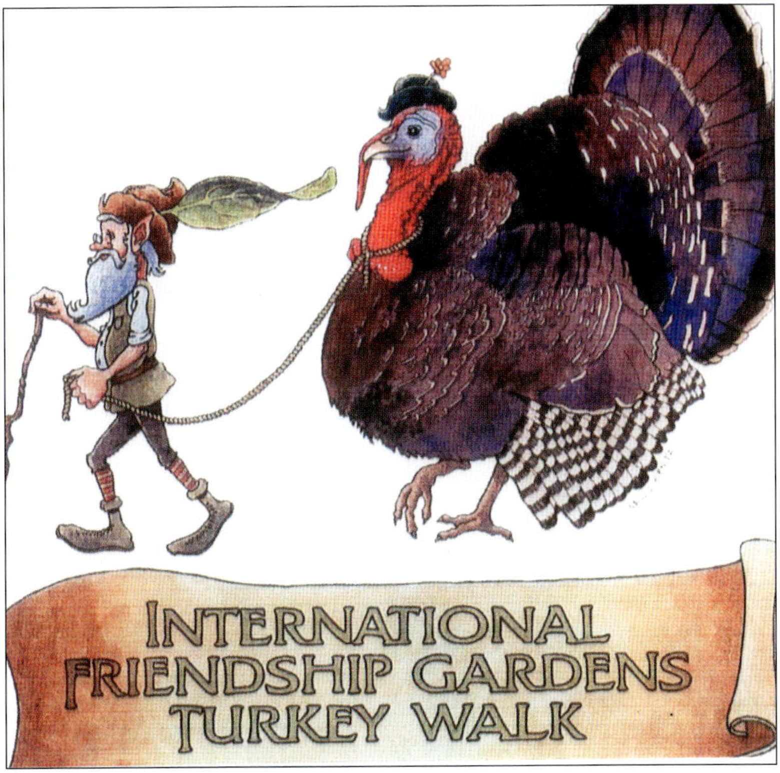

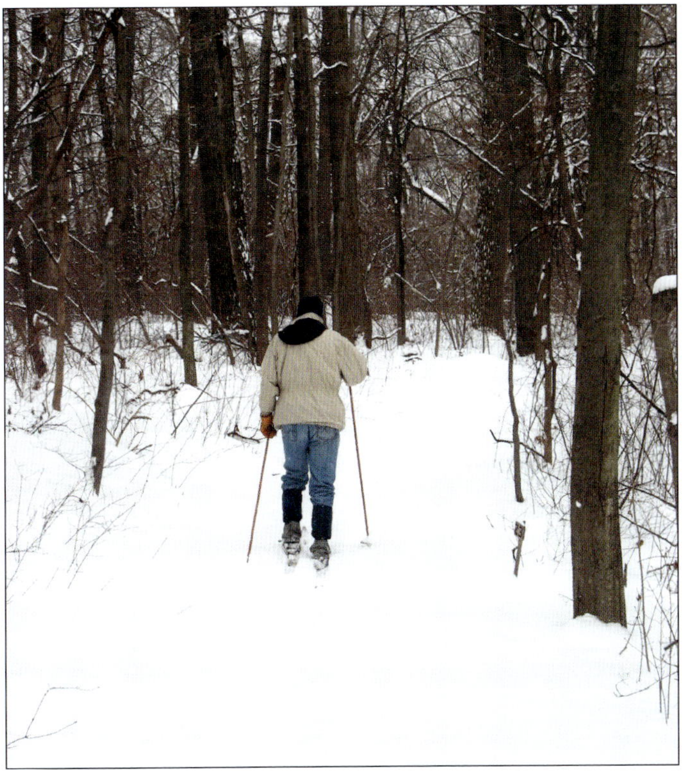

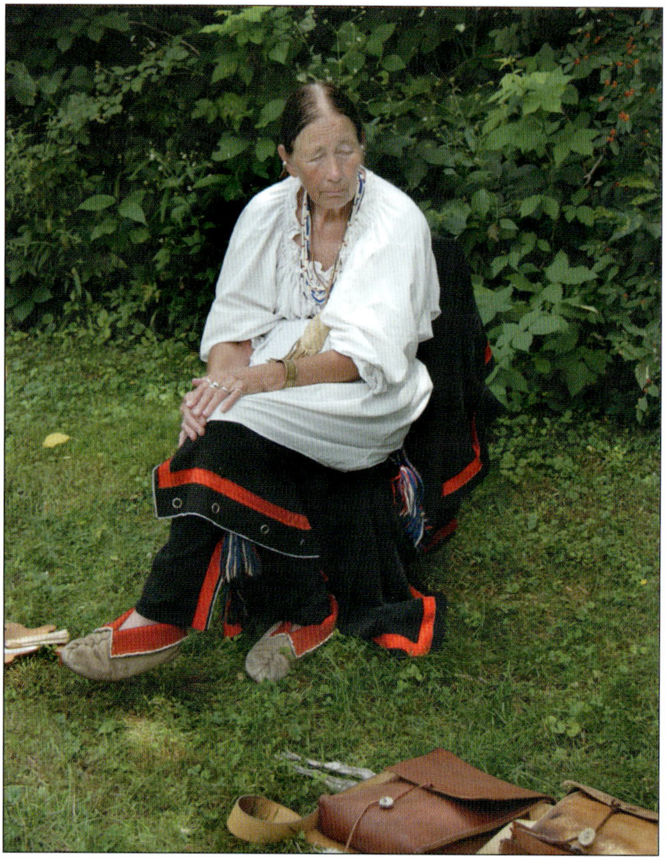

Native American Heritage Days have special significance in the Trail Creek region, where Potawatomi are known to have gathered more than 200 years ago. At left, Judy Judge, a weaver and historic reenactor from Kouts, Indiana, is photographed gazing pensively at her collection of handmade artifacts. Cooking over an open fire is demonstrated by Jude Rakowski (below), who previously managed Chellberg Farm and Bailly Homestead for the Indiana Dunes National Lakeshore.

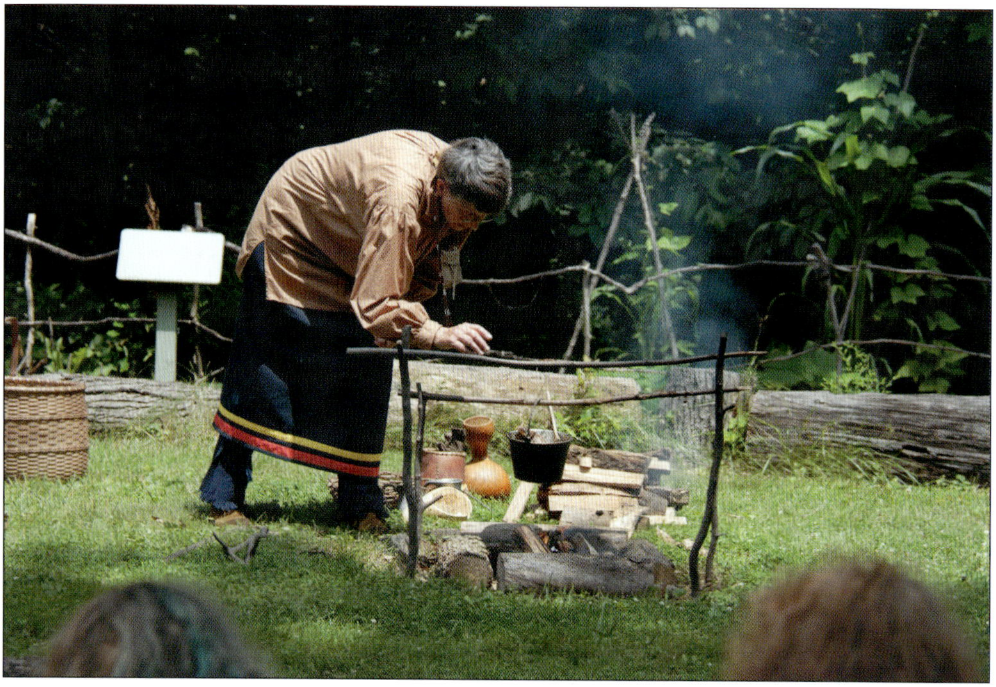

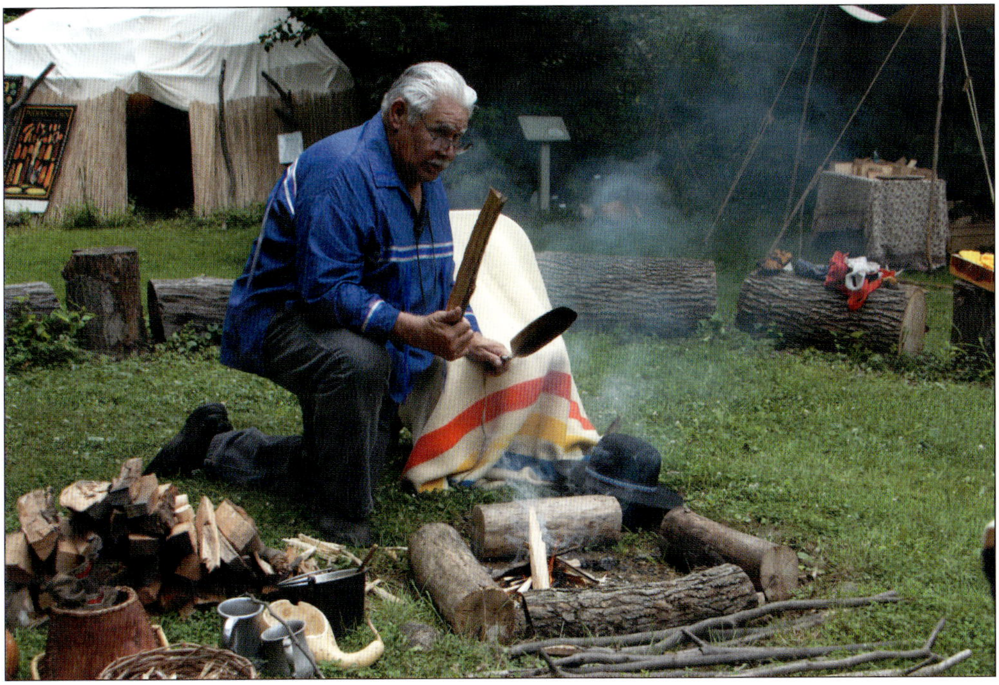

Henry Boulley (above), an Ojibwe elder who now lives in New Buffalo, Michigan, is known as a fire-keeper, which means he gets the job of starting fires and keeping them going. Handwoven baskets (below) are filled with berries, nuts, edible seeds, and mushrooms, evidence of nature's bounty in the wilderness. The Native American garden now has a council ring and fire pit, and a prairie garden has been developed, using native plant materials.

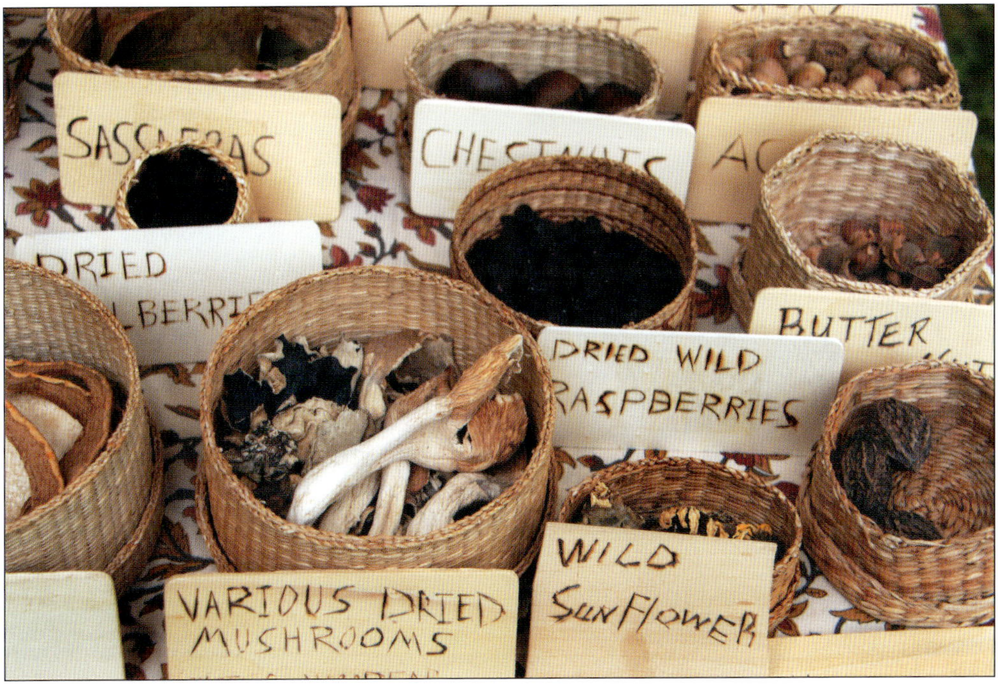

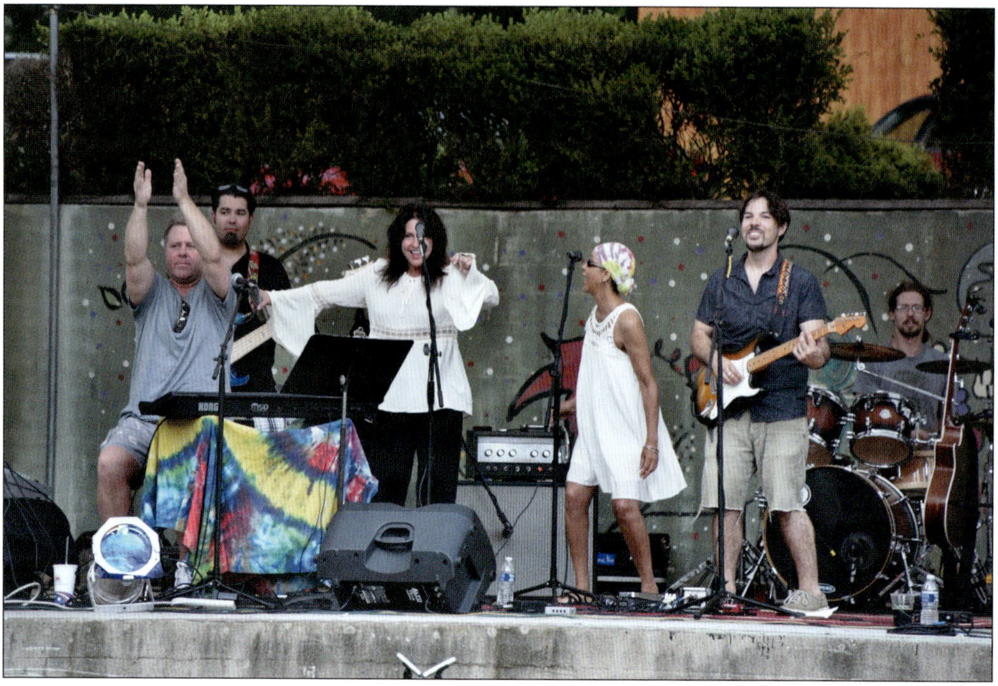

The Rootsdown Band (above) provided an evening of musical entertainment for many families, including Bobby Roberts and his niece Anna Triska (below).

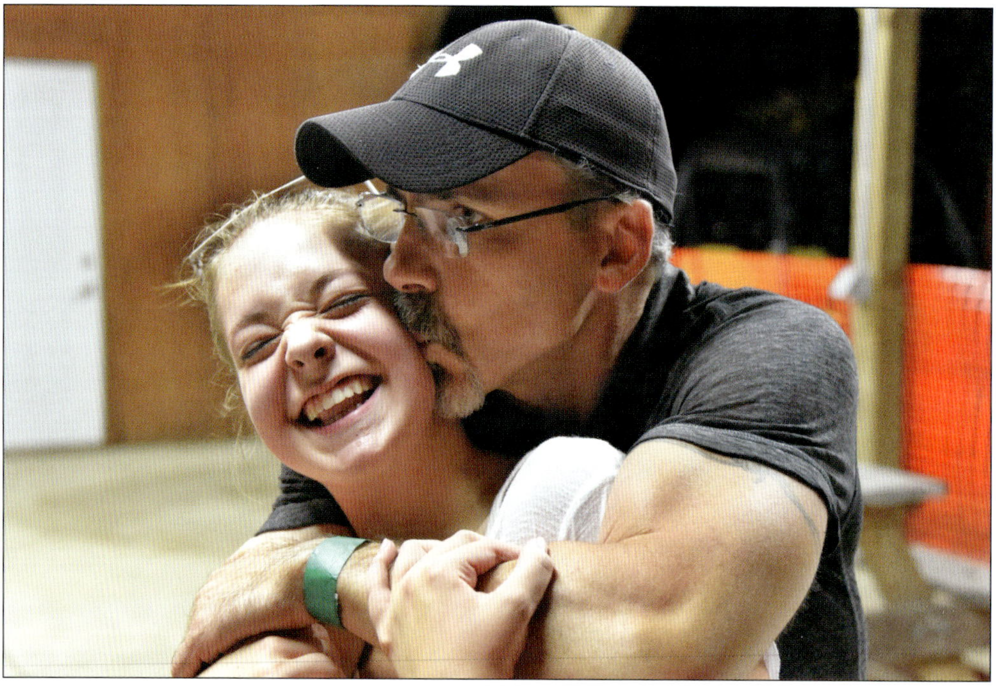

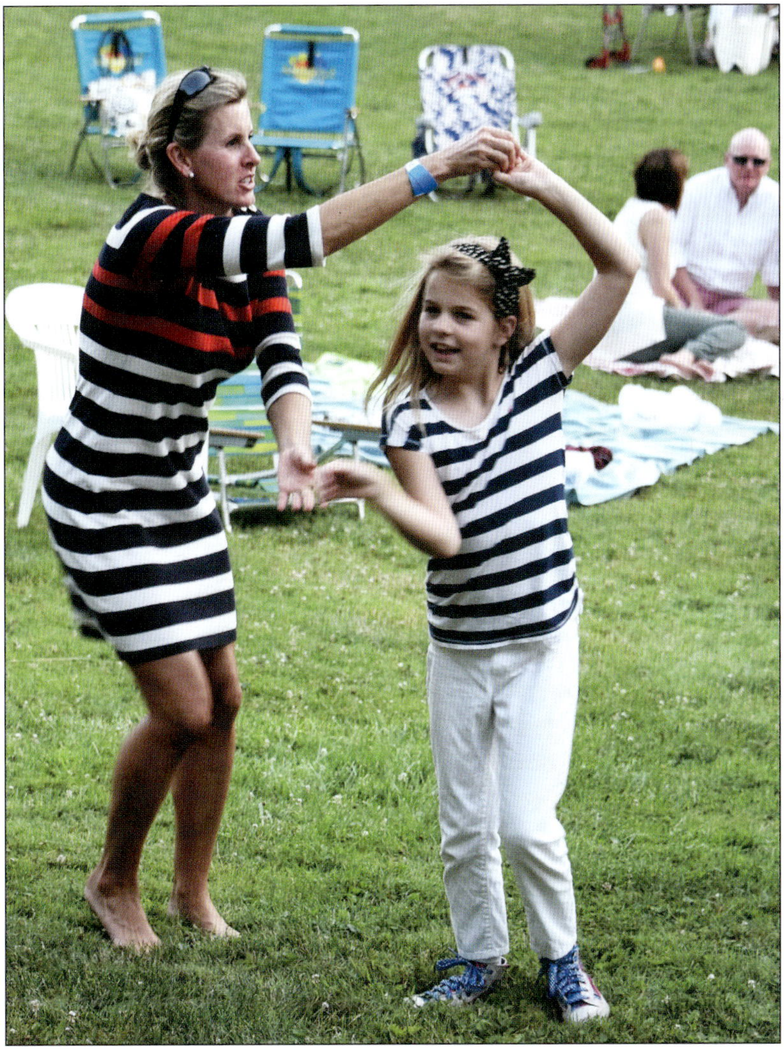

Concert Under the Stars brought dancers like Jeanie Shuck and her daughter Ellie Shuck to their feet (right). Others (below) spread their blankets and lawn chairs and enjoyed listening to the music.

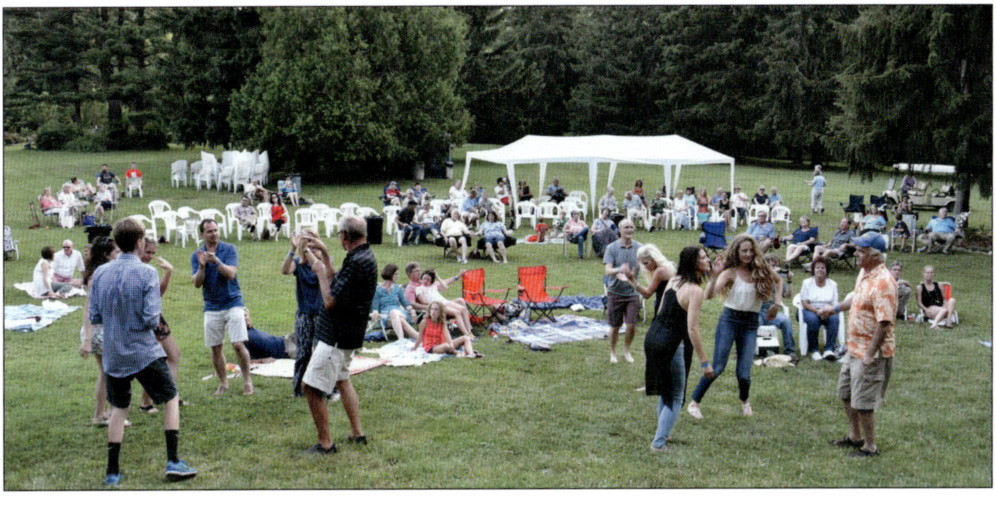

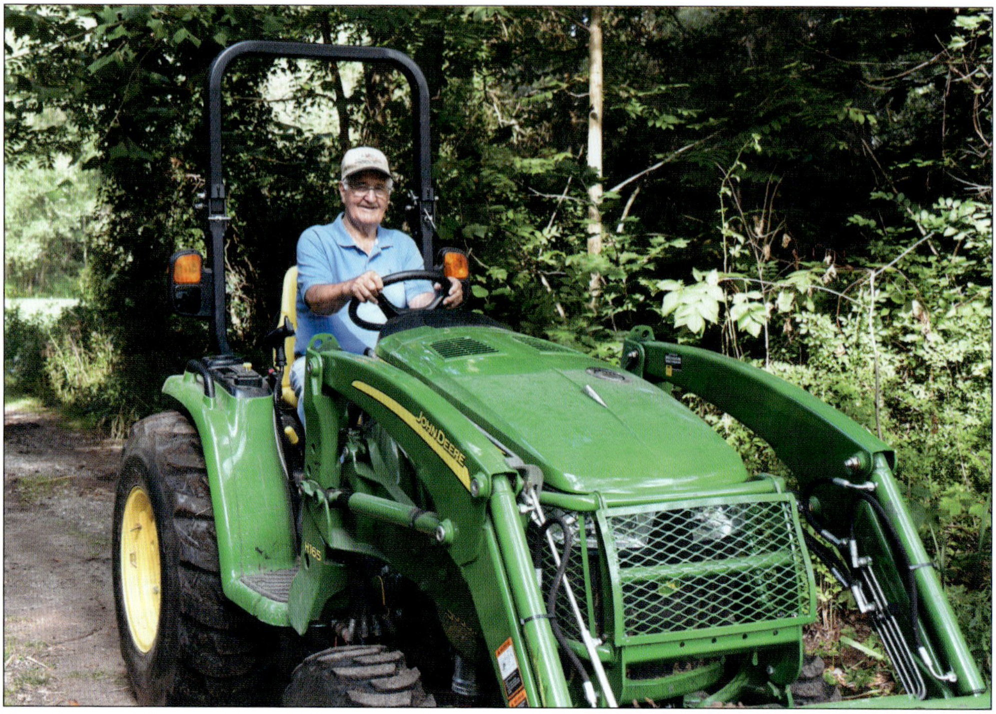

For the past 20 years, Ronald Bensz and his tractor have been crucial to the groundskeeping at Friendship Botanic Gardens. He personally cleared the path to open up the new entrance on US Highway 12. Bensz was a founder of B&E Marine and the Michigan City In-Water Boat Show. He is an ardent preservationist and community activist, with a special interest in the waterfront.

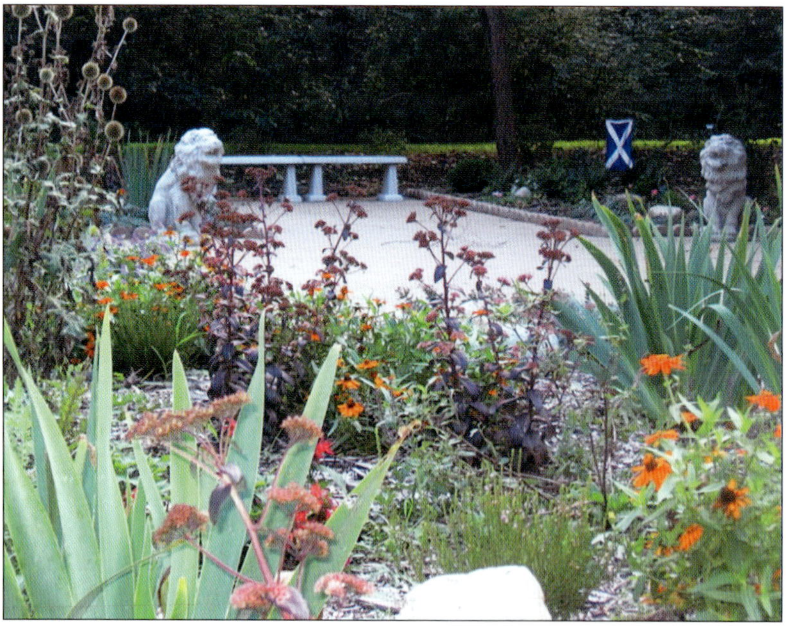

The guardian lions are now placed in the Scottish gardens, restored by Dr. Gene and Romona Hay in memory of their daughter Susan Hay Hemminger, who died in 1993 at the age of 34. Hemminger was a lawyer and Superior Court judge, the first woman to be elected judge in the history of La Porte County.

The German gardens, under the direction of Rev. David Albertin, are planted around a small pond.

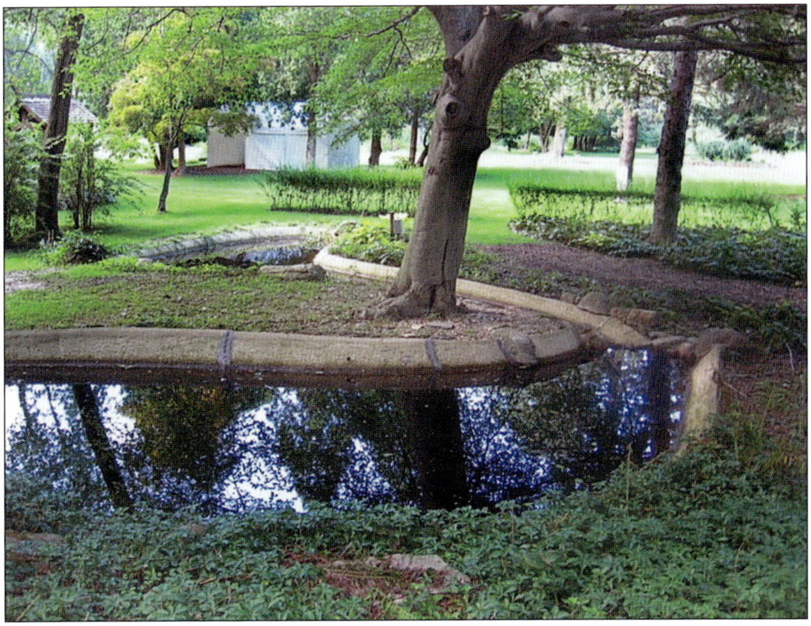

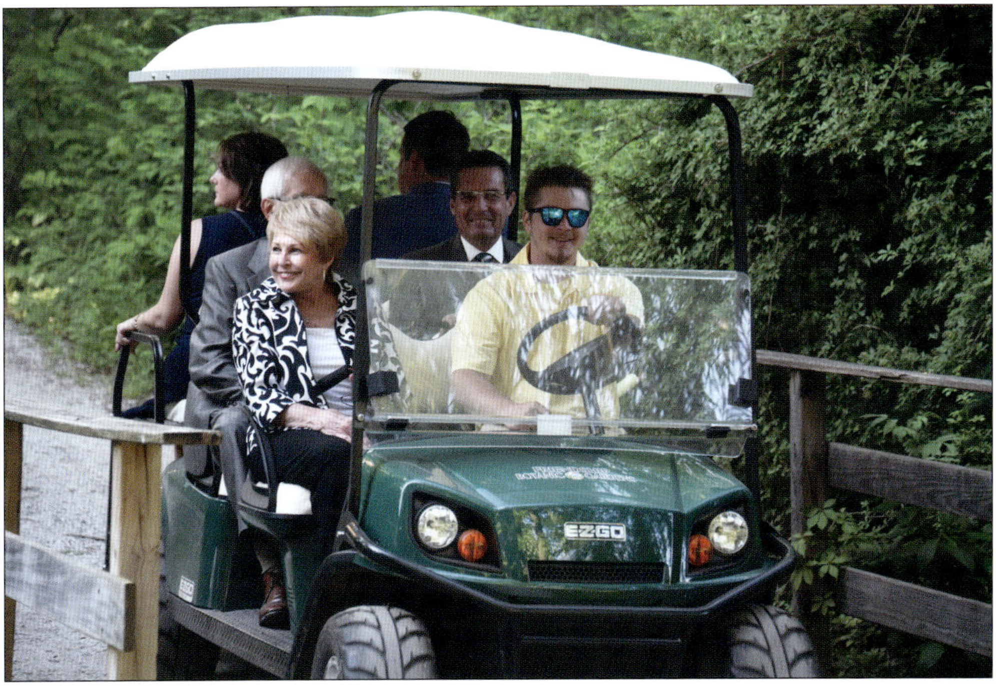

Brian Bentley (right front) transports a group of party-goers to the annual fundraising event in the gardens. Beside him is Judy Kroczek, an avid gardener and hybridizer of daylilies. She and her husband, Dr. Stephen Kroczek (seated behind her), have donated truckloads of hostas and daylilies to Friendship Botanic Gardens. The auctioneer for the evening is pictured at center; other passengers are unidentified. The new people-moving vehicles were donated by Barker Welfare Foundation.

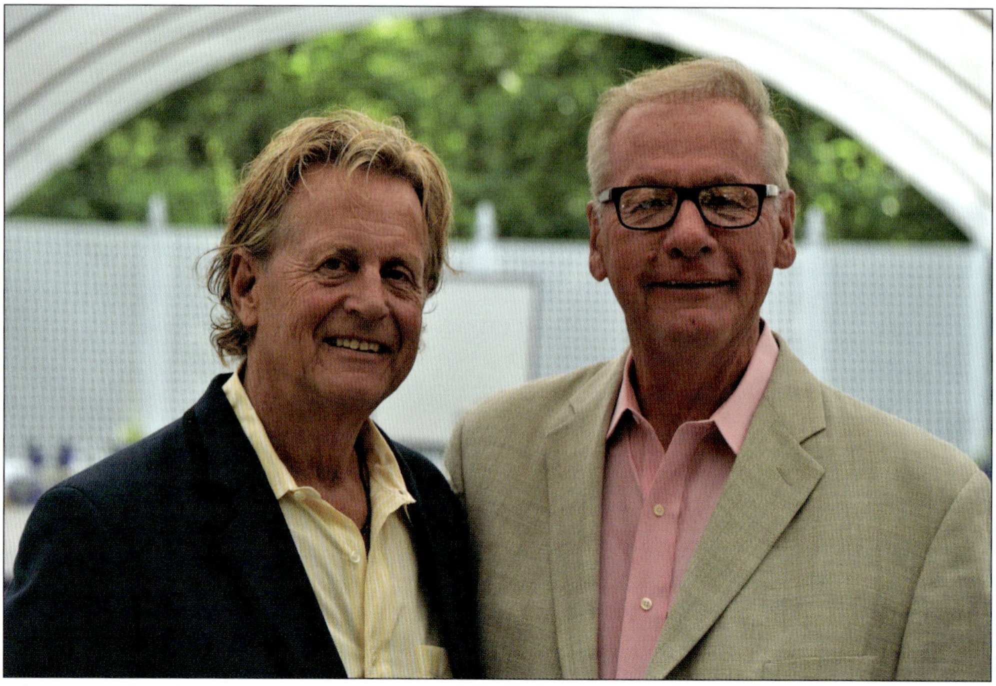

John Leinweber (above, right), board president of Friendship Botanic Gardens, is pictured at the annual fundraiser with his partner of 39 years, James Laughlin. Leinweber's business enterprises included start-ups of four document imaging companies. Laughlin is retired from his position in the health care industry. The couple has a historic home in Long Beach, Firefly horse farm in La Porte, Briar Leaf Golf Course, and a home in Florida. Below, a gala fundraising event draws a large crowd.

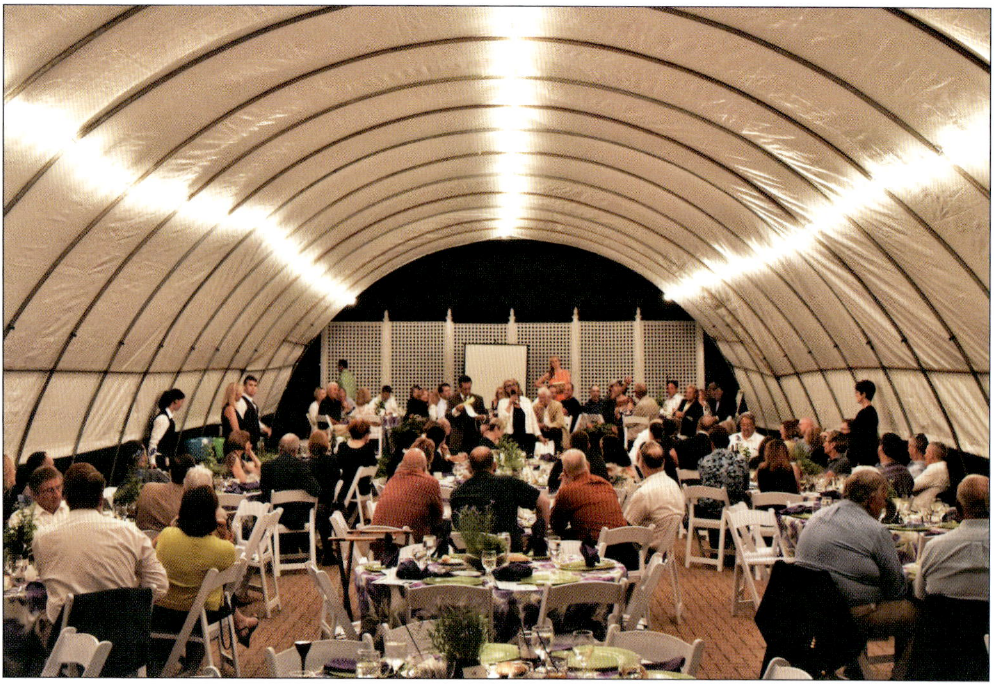

Vintage light posts placed along the trails were donated by Michigan City Redevelopment Commission, electrical work funded by NIPSCO and Marquiss Electric, Inc. The park benches were provided by Michigan City Park Department. The light posts and park benches were restored by Friendship Botanic Gardens.

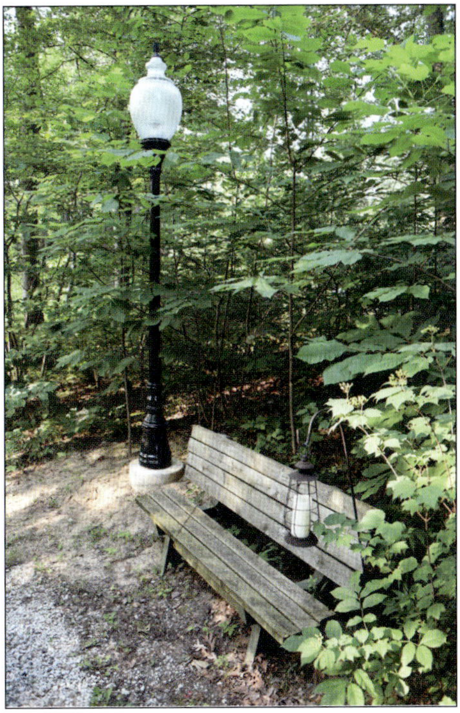

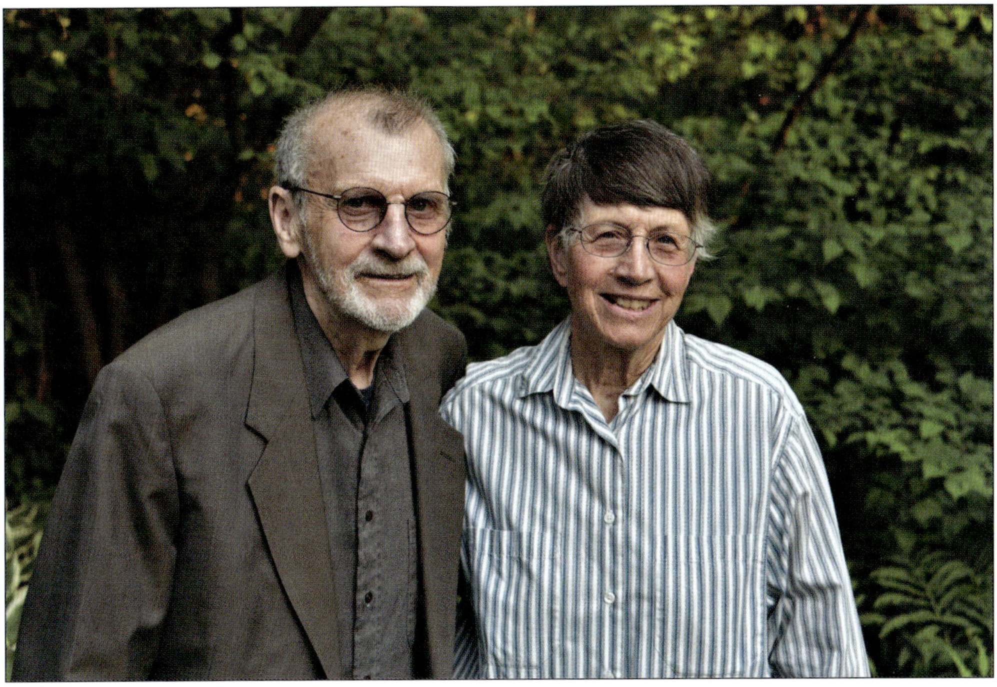

Alex Rakowski (left) and Jude Rakowski (right), residents of Long Beach, Indiana, were among the volunteers honored at the annual dinner. He is a local historian and archivist, retired as a counseling psychologist from Chicago State University. She is a retired park ranger from the National Park Service and a tour guide for the Audubon Society. Both are full-time volunteers at Friendship Botanic Gardens.

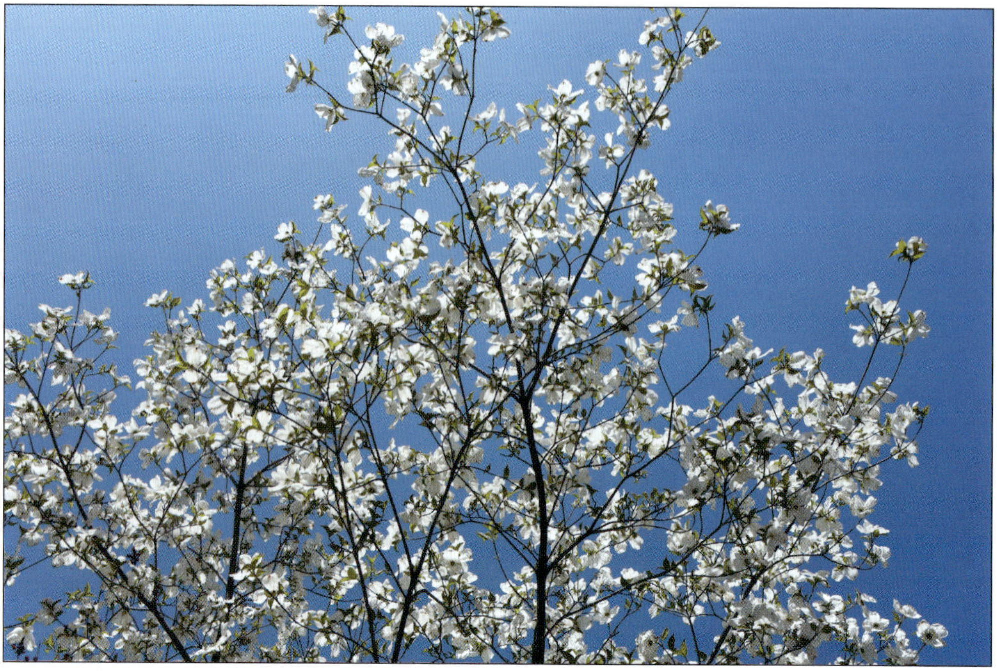

Dogwood trees announce the arrival of springtime at Friendship Botanic Gardens.

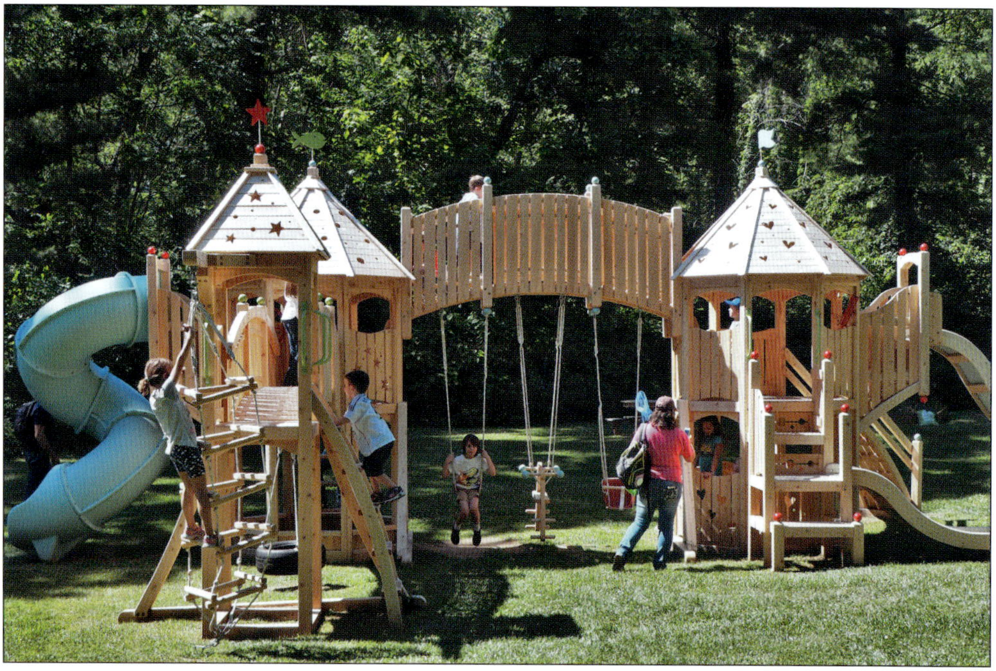

The ArcelorMittal Children's Garden has recently been outfitted with a remarkable piece of playground equipment, donated by the ArcelorMittal Corporation. The fortress-like structure includes swings, slides, and tunnels, with plenty of opportunity for exercise and adventure. It is made entirely of American cedar, good for at least 20 years and blending neatly into its wooded setting. John Leinweber and Kevin Egan, observing the great popularity of children's activities at Friendship Botanic Gardens, found this piece in Maine and had it shipped to Indiana.

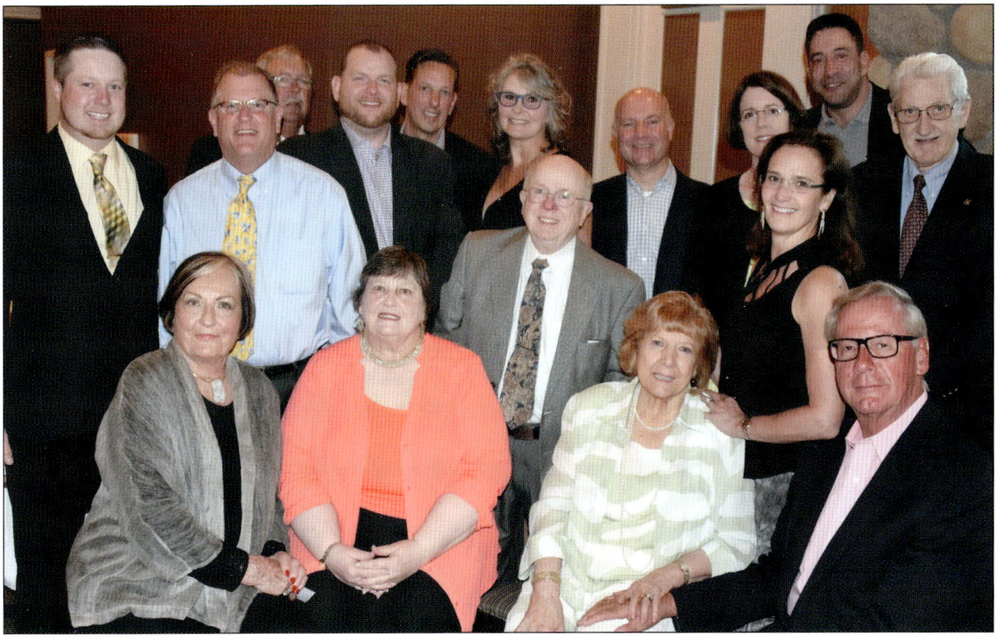

The Friendship Botanic Gardens Board of Directors consists of, from left to right, (seated) Dr. Rima Binder, vice-president; Mary Fox, attorney, treasurer; Jeanne Houck, director emerita; and John Leinweber, president; (standing) Richard Houck III; Dr. Kevin Egan; Steve Kahn, vice-president; Craig Phillips, Michigan City director of planning; Alexander de Hilster; Rubia Jasinevicius, secretary; Dr. Richard Houck, president emeritus; unidentified; Cathy Kearney; Leslie Samuelson; Don Dixon; and Olief Olsaker.

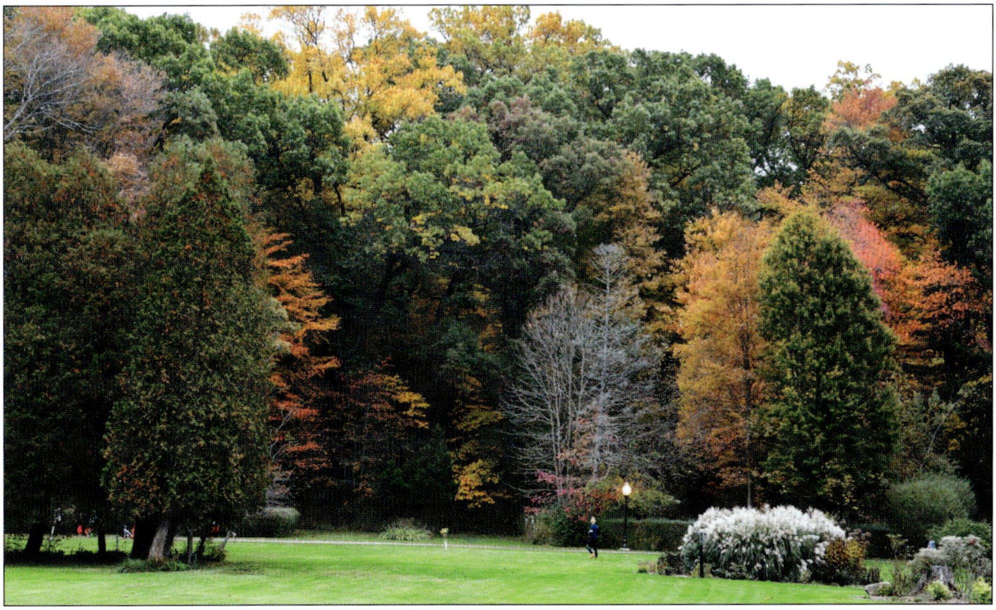

Autumn leaves bring a spectacular display of color to the many types of trees at Friendship Botanic Gardens.

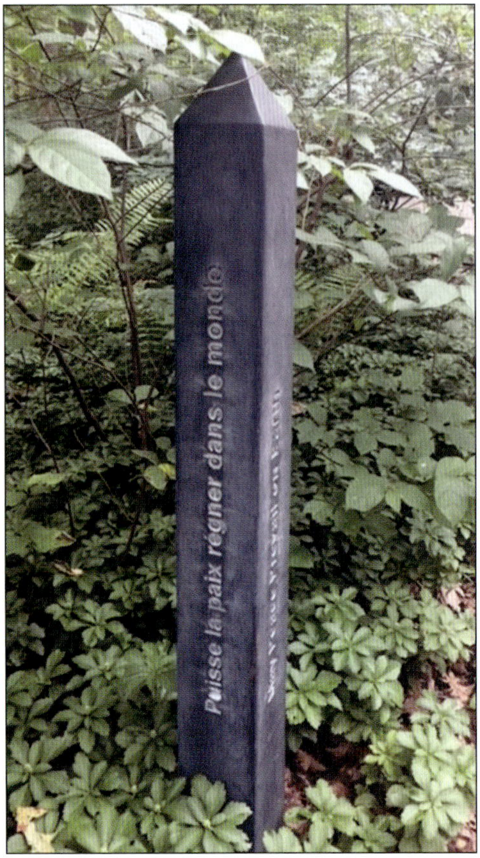

The Peace Monument was donated to the gardens by Ann Dahm, then president of the Michigan City Chamber of Commerce. She was acquainted with sculptor Olivia Smith, a former resident of New Orleans who now lives in New York City. The monument is a four-foot obelisk inscribed with the universal prayer, "May peace prevail on earth" in four languages—English, French, Spanish, and Arabic.

The Chicago sign also has a story. It is a brownstone carving ordered in 1914 for Chicago's Union Station and was to accompany a huge figure of Fr. Jacques Marquette. World War I disrupted this project. By the time the station was completed in 1925, the railroad no longer wanted the statue. It languished in the Polk Street freight house for 15 years and then was offered to Ed Paine, whose sister Gladys lived in Long Beach. She thought it perfect for Marquette Spring, near International Friendship Gardens. Virgil Stauffer agreed and had it shipped out but never managed to get it reassembled. Alex Rakowski rediscovered the fragments in 2001.

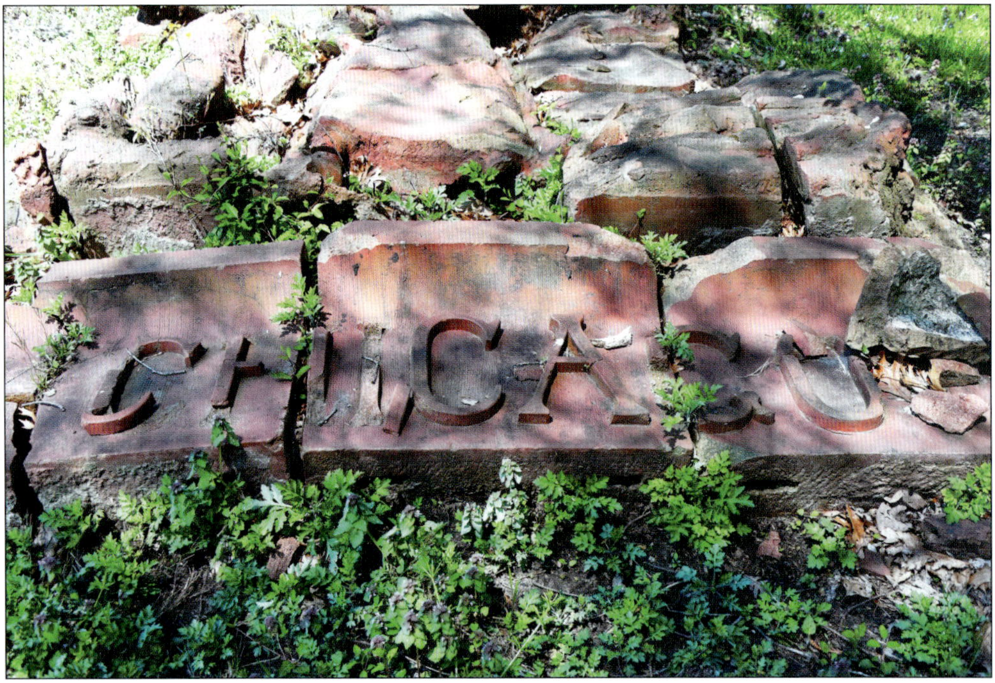

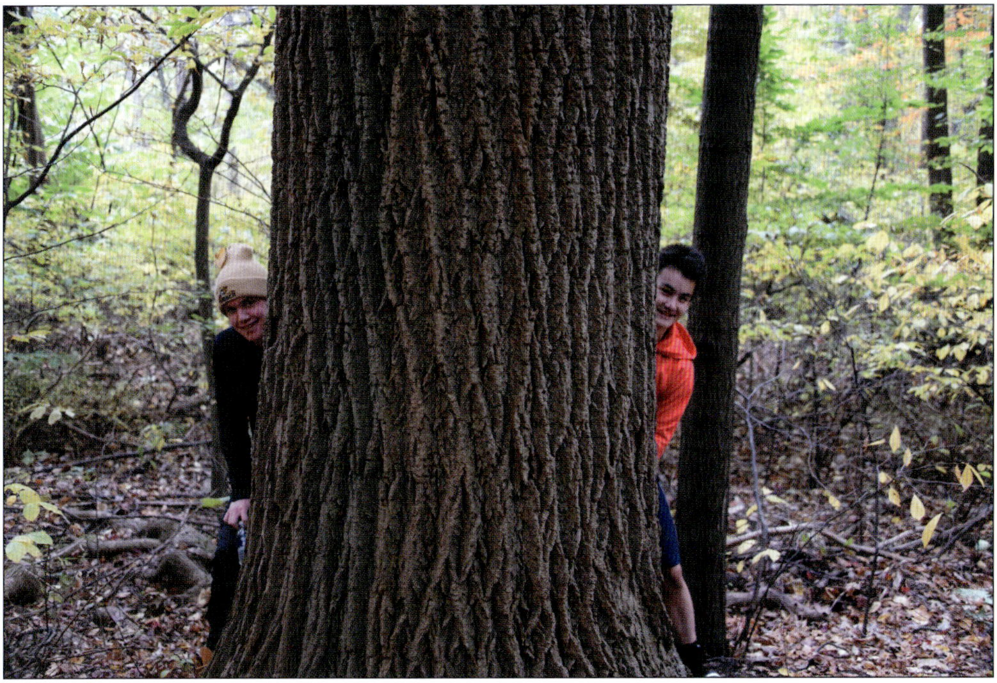

A huge tulip tree provides a tempting spot for games of hide-and-seek.

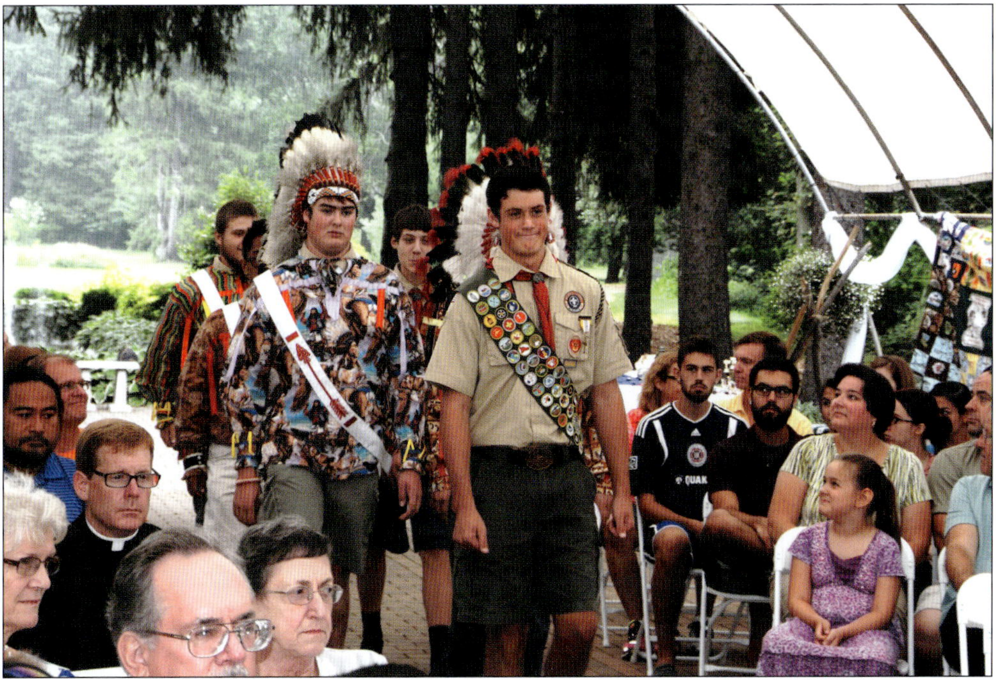

Friendship Botanic Gardens was selected as the setting for Joseph Gondek's Eagle Scout investiture ceremonies in August 2014. Several Native Americans participated in the event as well as officers of the Boy Scouts of America.

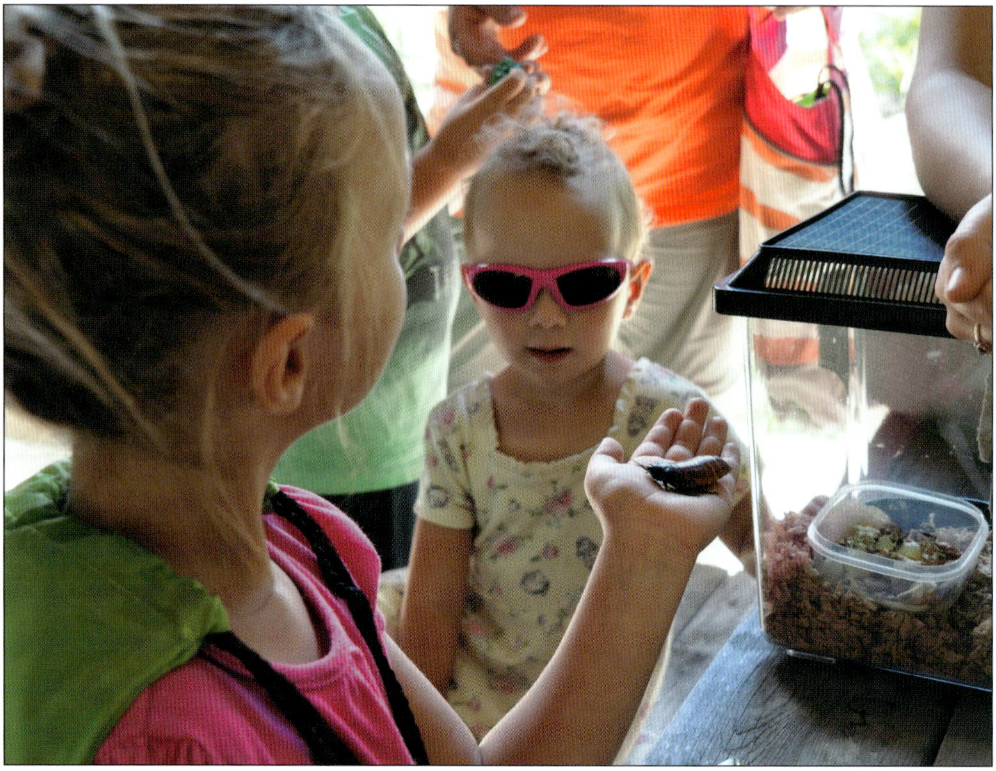

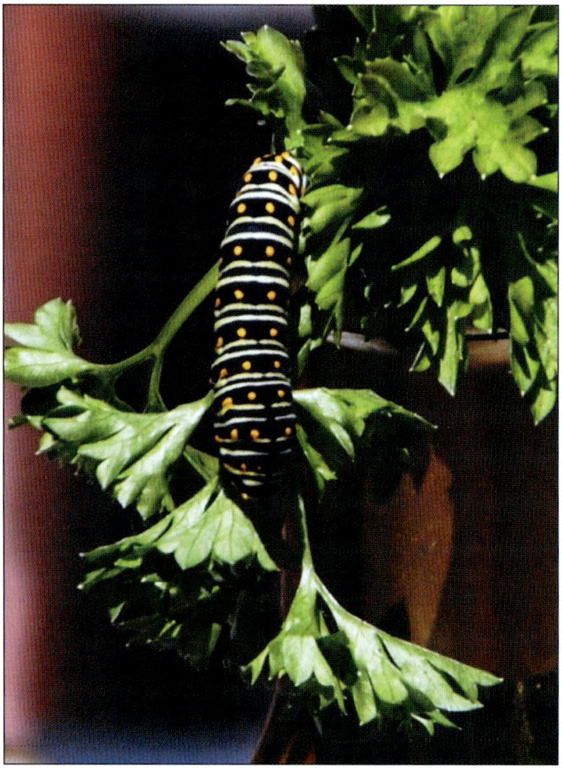

At Bug Safari (above), children learn how to handle insects and observe their behavior, which they evidently find intriguing. At left, a beautifully striped caterpillar is photographed during his lunchtime.

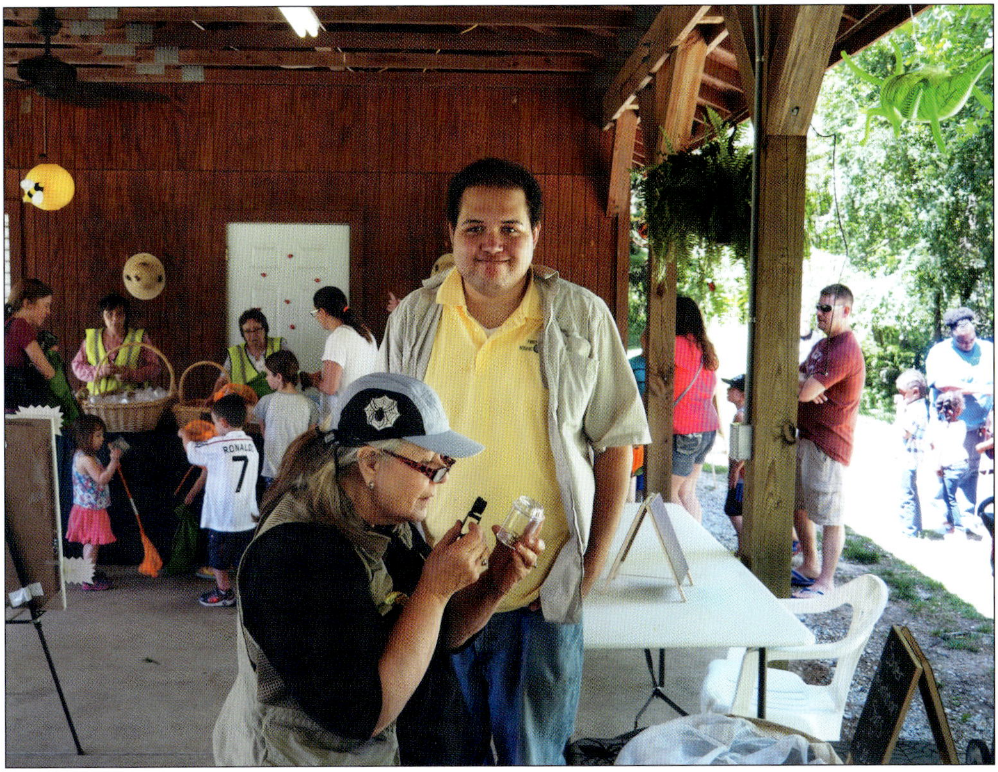

Ronald Taylor (above, center), communications director at Friendship Botanic Gardens, is fascinated by insects and finds they have endless potential for educational projects. As the founder of Bug Safari, Taylor says, "These exquisite creatures open up a wonderful little world for children to enjoy." The Madagascar cockroach (pictured below, climbing on sunglasses) is a playful insect, often selected as a pet. "It makes a hissing sound," Taylor says, "as a self-protective device or to communicate."

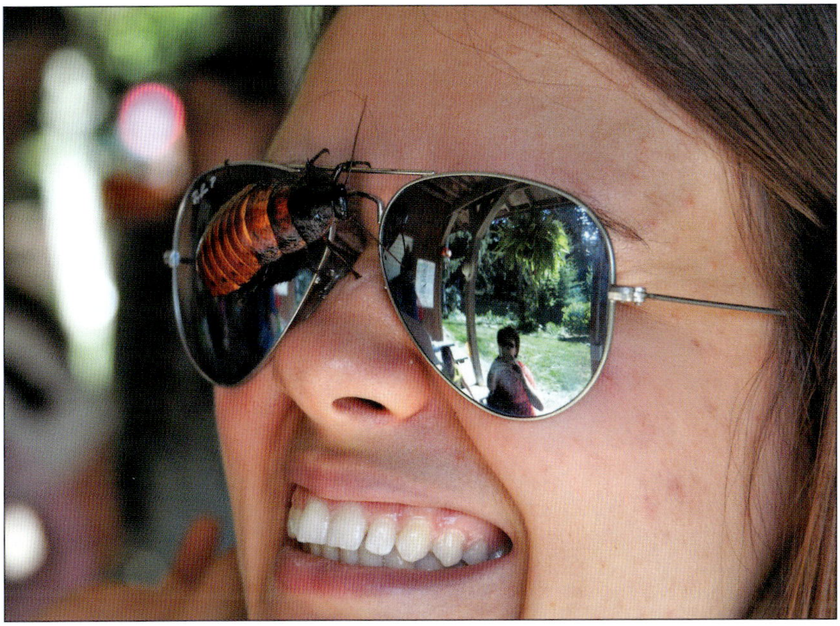

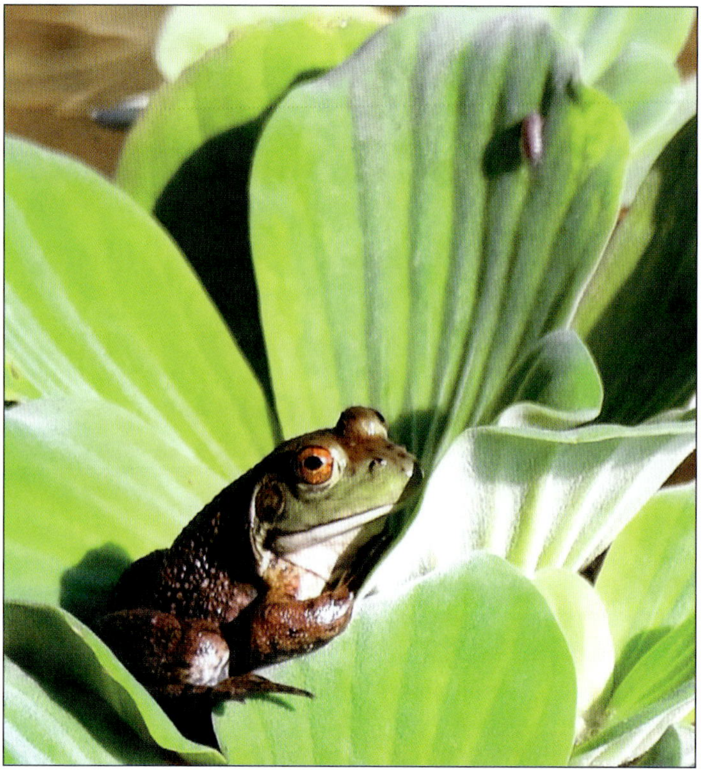

Friendship Botanic Gardens offer natural habitats for many varieties of wildlife, including a frog (left). Audubon Society members have spotted 40 types of birds, including woodpeckers, warblers, hawks, an indigo bunting, Baltimore oriole, barn swallow, American goldfinch, and flocks of Canada geese. The bronze and copper sculpture of the great blue heron (below), a permanent resident, was donated by Linda Schwartz and Bruce Gershenzon.

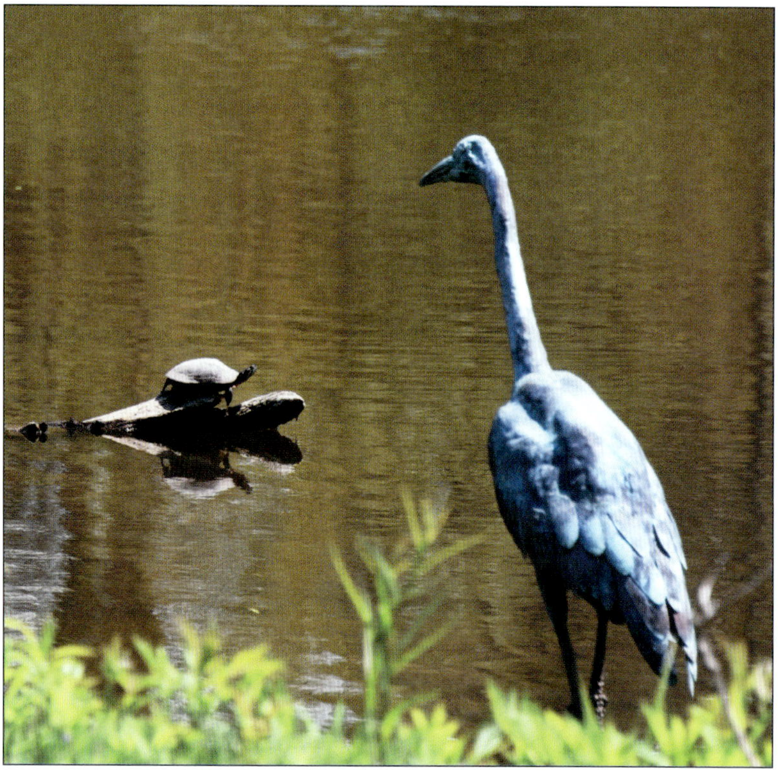

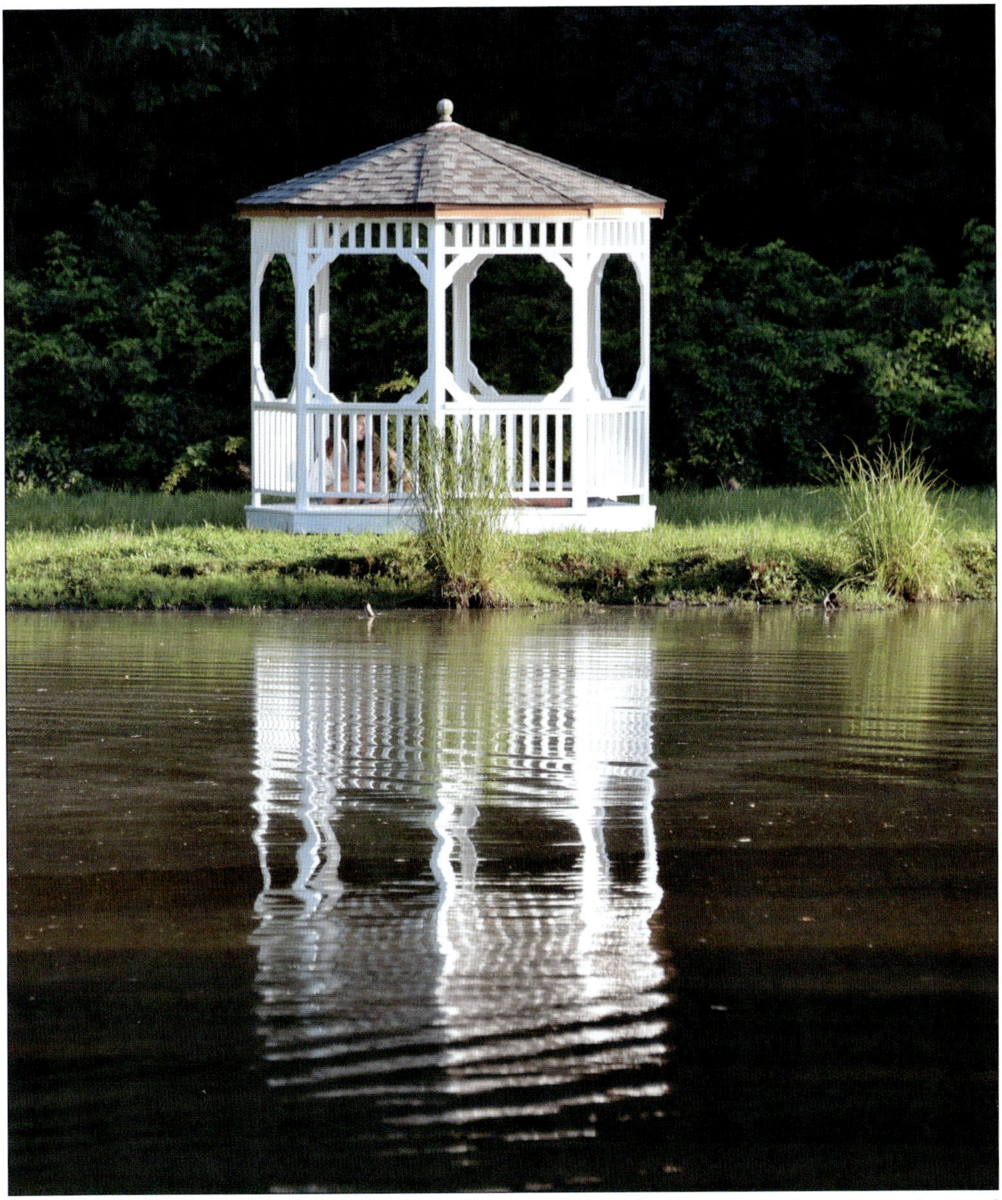

A gazebo donated by the Cypra family in honor of Ken Cypra, former board member and philanthropist, has been built on the island of Lake Lucerne. It has been designated the Cancer Survivors Garden of Love and Hope, in keeping with the spirit of dedication to special causes. A Juvenile Diabetes garden exists, and plans are under way for a new garden honoring veterans of foreign wars.